Understanding Radioactive Waste

Fourth Edition

Understanding Radioactive Waste

Fourth Edition

Raymond L. Murray, Professor Emeritus
Nuclear Engineering Department
North Carolina State University
Raleigh, North Carolina

Edited by
Judith A. Powell
Battelle Pacific Northwest Laboratories

 BATTELLE PRESS
Columbus • Richland

The first edition of this book was prepared by Battelle's Pacific Northwest Laboratory for the U.S. Department of Energy.

Cover Art Courtesy of Chem-Nuclear Systems, Inc.

Library of Congress Cataloging-in-Publication Data

Murray, Raymond LeRoy, 1920-
    Understanding radioactive waste / Raymond L. Murray. -- 4th ed.
    p.   cm.
    Includes bibliographical references and index.
    ISBN 0-935470-79-4 : $12.50
    1. Radioactive wastes.   2. Radioactive waste disposal.
I. Powell, Judith Ahrens.   II. Title.
TD898.M87   1994   94-9995
363.72'89--dc20   CIP

Additional copies may be ordered through Battelle Press, 505 King Avenue, Columbus, Ohio 43201, U.S.A., (614) 424-6393 or 1-800-451-3543.

To seekers of sound solutions

Acknowledgments

This book seeks to present facts about all aspects of radioactive wastes in a simple, clear and unbiased manner. The information is intended for students and other interested or concerned members of the public.

Because many diverse fields are involved and the book is in its fourth edition, advice and assistance from many persons have been needed.

Special thanks are due John V. Robinson for his conception of the book and continued advice and encouragement. Thomas D. Chikalla and L. Donald Williams of Pacific Northwest Laboratory have been uniformly supportive.

Valuable suggestions have been provided by M. Ray Seitz, Edwin L. Fankhauser, Donald Fankhauser, Scott Nealey, Douglas Robinson, Margaret N. Maxey, David W. Levy, Frank K. Pittman, Eugene A. Eschbach, Lawrence J. Smith, Stanley M. Nealey, John T. Suchy, Harry Babad, Kenneth J. Schneider, Orville F. Hill, Harold H. Hollis, Dennis K. Kried, Elizabeth Reid Murray, Bonnie Berk, Thomas Overcast, Janie Shaheen, Richard Williamson, Faith Brennemon, Melvin Shupe, Paul Dunigan, Jack L. McElroy, Harry C. Burkholder, Ted Wolff, and Steven C. Slate.

Illustrations have been prepared by Van Webb, Ann Kennedy, Valerie Stott, and Luann Salzillo. Thanks are due Sandra K. Blakley, Janet K. Tarantino, Donna Gleich, Darlene E. Whyte, Leslie L. Hughes, and Joseph E. Sheldrick for important contributions to the production of the book.

The author and editor appreciate very much the generous assistance of these people. In addition, we wish to thank each other for a rewarding collaboration, marked by stimulating discussion, excellent cooperation, and mutual understanding despite our never having met. We believe that the book's favorable reception owes much to that happy association and a shared philosophy regarding honesty and clarity in writing.

Raymond L. Murray
Raleigh, North Carolina

Judith A. Powell
Richland, Washington

Contents

Illustrations

Tables

CHAPTER 1
Questions and Concerns
About Wastes

Energy and the Environment

Today the world faces two major problems related to energy and the environment: pollution and potential scarcity.

The environmental movement of the 1960s called attention to the growing problem of pollution to land, air, and water both in developed and developing nations. Concern increased around the world about the release of chemicals. In the 1970s and 1980s the problem of chemical waste became prominent. It is clear that large amounts of hazardous chemicals were stored or discarded with inadequate precautions. The *Exxon Valdez* oil spill in Alaskan waters in 1989 caused extensive environmental damage. People became increasingly aware also of radioactive wastes, those that emit radiation as they break up or "decay."

The Arab oil boycott of 1973 focused attention on the potential scarcity. Sources of oil are in the hands of unstable or vulnerable countries. Costs of petroleum increased significantly, contributing to economic difficulties in both advanced and emerging countries. The Persian Gulf War of 1991 was a response by the United Nations to a request from Kuwait, an important oil-producing country. Ironically, that war resulted in Iraq's committing ecoterrorism by burning and dumping 250 million gallons of oil, severely polluting land, air, and water.

U.S. energy policy is to minimize oil vulnerability, to improve efficiency of energy use, to stimulate economic growth, to help developing nations, to encourage alternative sources to fossil fuels, and to protect the public and the environment.* Nuclear power is viewed as one choice to meet national needs, along with natural gas, coal, and solar energy.

The Earth Summit Conference in Rio de Janeiro in June 1992 called attention to the challenging global problems of resource transfer, sustained development, and environmental protection.

---

*Energy Policy Act of 1992, Public Law 102-486, October 24, 1992.

Questions About Nuclear Energy

Although people recognize the need for various energy sources, they often raise the question about nuclear energy, "What is being done with radioactive wastes?" The federal government and the nuclear industry have stated that such wastes were known to be dangerous ever since they first were generated in large quantities in World War II, and that special care has been taken over the years to protect the public. Observers note, however, that decisions about final disposal of wastes have had a lower priority than production in the overall nuclear development. Examples of poor strategy and uncertainty in the waste management program are cited. The search for suitable disposal sites has stimulated public concern. News of disagreement among different branches and levels of government has led many to conclude that we do not know what to do with radioactive wastes.

The question "Is nuclear power safe?" is also raised frequently. People are aware of the fearful effects of the atom bomb. They know that a nuclear weapon is not the same as a nuclear reactor, but they tend to associate the two and are uneasy because both involve fission and radioactivity. Many people think of all radiation as mysterious and lethal. When these ideas are combined with Murphy's law, "If anything can go wrong, it will," it is easy to see why many people are uncomfortable, worried about, or frightened by continued or expanded use of nuclear energy. Most people are aware that there are few fatalities due to the use of nuclear energy. The Chernobyl reactor accident of 1986, however, heightened fear of nuclear power throughout the world. People are aware of greater dangers, but feel that in those situations they are in personal control of their safety. The nuclear hazard is less acceptable than other, more familiar, risks such as riding in automobiles.

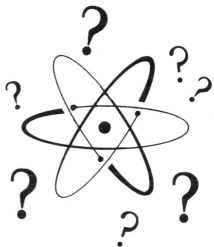

Scientists are expressing alarm about damage to forests and inland waters and a possible global temperature rise due to emissions from burning fossil fuels. For this reason, the question also is heard, "Does nuclear power help solve the problems of atmospheric pollution, acid rain, and the greenhouse effect?"

The Need for Information

Approval for new nuclear electric plants and continued use of existing ones may depend on satisfactory demonstration of safe waste disposal. In making decisions about power plants it is important that citizens and lawmakers alike know the nature of the waste problem and be able to distinguish opinions, feelings, and myth from the facts.

Unfortunately, too little public information on the subject of radioactive wastes is useful. Much of that said or written is rhetoric intended either to frighten or soothe. Debaters exaggerate to try to convince rather than inform. Polarization is such that neither side of the argument is credible to the average person.

The public is often confused by conflicting statements about nuclear energy and the waste problem. Scientists of presumed equal qualifications, for example, Nobel Prize winners, are seen taking opposite stands. There is a wealth of technical literature, but it is written for use by scientists and engineers familiar with the technical terms and background. Reports on plans and progress are in formal governmental language, which is often hard to translate into ordinary English.

This writer hopes to accomplish several things: (a) explain clearly the origin and nature of nuclear by-products, (b) provide facts and figures about nuclear wastes and the actions being planned on a national basis, (c) provide perspective on the safety of waste isolation systems, and (d) distinguish knowledge from opinion whenever possible, in an unbiased and candid manner. The author rejects exaggerated statements about the waste problem at both poles of the debate—assertions by proponents that it is merely a matter of politics, by opponents that a technical solution is impossible.

Understanding and Decisions

The most important premise behind this book is that an informed public will make the best decisions. We intend to help the reader understand such nuclear topics as uranium, radioactivity, radiation, and fission, along with the role of materials, chemical processes, and geology in the treatment and long-term handling of wastes. We will touch on people's knowledge and attitudes about radioactive wastes.

The nuclear debate may be characteristic of our times. Some say, "Government bureaucracy and industry try to go ahead with no interference by the public, but fortunately there are public interest groups to help the people speak out." Others counter, "We have dedicated leaders seeking to serve the public most effectively and economically, versus a small number of activists who would like to see power decentralized and the present system dismantled." Perhaps there is some truth in both views. The author believes that all thinking citizens have some interest in and concern about the waste problem and that a thorough explanation will be useful.

Some questions we hope to answer to the reader's satisfaction are:

1. What is radioactivity?
2. How are radioactivity and radiation related?
3. What are the effects of radiation on living beings?
4. How can we be protected from radiation?
5. In what ways is radiation useful or beneficial?
6. What are radioactive wastes?
7. Where do most of the radioactive wastes come from? How much do we have on hand and how much is produced each year?
8. What is the difference between high-level and low-level wastes?
9. How are nuclear wastes similar to or different from poisons from chemical processes? From emissions of electrical power plants using coal or oil?
10. What are nuclear power plants doing with their used fuel?
11. Why do we not yet have a place to dispose of spent fuel? When will a repository be available?
12. How are used nuclear reactor fuel and processed nuclear wastes transported? What precautions are taken to prevent accidents and to protect the public?
13. Should people be informed of waste shipments?
14. What are the best ways to dispose of wastes?
15. What laws and regulations apply to the management of wastes?
16. What organizations are responsible for handling wastes?
17. How much space is needed to hold wastes? How far from public areas should disposal sites be located?
18. What kinds of rock are suitable for waste disposal? What do we need to know about rock formations?
19. How much does it cost to dispose of wastes? Who pays for it?
20. Is it safe to dispose of waste as planned?
21. What will be done with nuclear weapons materials as disarmament proceeds?
22. Is the problem of nuclear waste disposal overemphasized compared with other national problems?

The Sciences Help Explain

What one should learn so as to understand both the questions and the answers depends on a person's attitude. No information is needed if one says either "I am satisfied that those in government and industry will make good decisions" or "I am convinced that nuclear power is unsafe and must be abandoned at once." At another extreme, if one says, "I must have the same background as scientists and engineers who are advising on wastes," years of specialized college training would be required.

We take a middle ground and assume that the reader wishes to gain enough technical information to be able to think clearly about the subject and discuss it rationally. We believe it is helpful to know some atomic and nuclear science, to have been exposed to a little of the history of nuclear energy development, and to have some appreciation of what a nuclear reactor is and does. This background helps in learning about radioactive wastes and their management, and in understanding the issues that still remain to be resolved before the waste problem is put to rest.

Some basic science background is needed because many subjects bear on nuclear wastes—general science, chemistry, physics, biology, and earth science. Let us see how these sciences come into play:

- Wastes consist of many chemicals, composed of about half the 109 elements in the periodic table. Chemistry tells us how well we can remove certain hazardous substances, what form the wastes are in, what they will mix well with, how easily they are dissolved, and how fast they move through air, water, and earth.
- Physics describes properties of the nucleus of the atom, including how many types of radioactive atoms there are, how long they last, and what radiation they emit.
- Biology explains how radioactive materials may be brought to man through food chains involving plants and animals, which organs of the human body have an affinity for certain materials, and how radiation damages cells and tissues.
- Earth science describes the features of rocks and soil in which wastes may be placed, explains how water is transferred through the ground, and identifies the types of medium and best locations for long-term waste storage.
- Meteorology tells where airborne nuclear wastes might go and what might be the effects of emissions of other energy sources besides nuclear.

In the next several chapters we review our knowledge of the atom, the periodic table of elements, simple chemical reactions, the nature of isotopes, radioactivity, and radiation. Special attention is given to effects of radiation and ways to protect human beings. We need such background for an understanding of radioactive waste.

CHAPTER 2
Atoms and Chemistry

Some Distinctions: Atomic and Nuclear

The scientific basis of radioactive waste management involves two levels: *atomic* and *nuclear*. These words are often confused—they refer, however, to quite different realms of matter. Atomic processes are chemical in nature, involving electrons of atoms or molecules. Nuclear processes are very energetic ones involving the inner core of the atoms.

Thus chemistry reveals what treatment or processing of used (spent) fuel from nuclear reactors will separate the useful from the useless. It tells what wastes might combine well with and resist attack by acids and tells what metals will make good containers. It also indicates how to prevent waste-particle migration in the ground.

On the other hand, the subject of nuclear physics explains the processes of radioactivity, the behavior and effects of radiation, and the production of new species of material by neutron bombardment in a reactor.

To understand the radioactive waste problem, we need both atomic and nuclear concepts. Let's review, very briefly, the important facts as discovered in the last hundred years or so.

Our World of Atomic Chemistry

Recall from basic science courses that all matter is made up of a small number of different kinds of atoms, the chemical elements. The first and lightest of these is hydrogen, and the 92nd and heaviest natural element is uranium. There are 17 manmade elements that are heavier than uranium.

Chemical elements combine to form new substances called compounds. Some familiar processes, expressed as equations, are these:

burning of hydrogen:
    hydrogen + oxygen → water

preparation of table salt:
    sodium + chlorine → sodium chloride

rusting of iron:

iron + oxygen → iron oxide

burning of natural gas:

methane + oxygen → carbon dioxide + water

neutralization of an acid by an alkali (lye):

nitric acid + sodium hydroxide → sodium nitrate + water.

More complicated still are the reactions in living organisms, such as photosynthesis in plants and digestion in animals. The regularity with which chemical reactions take place led to the discovery of the atomic theory—that matter is composed of individual particles called atoms, which combine to form molecules. Atoms are extremely small. It would take a row of 10 million hydrogen atoms to span the head of a pin. One teaspoon of water contains an enormous number of molecules—around 200,000,000,000,000,000,000,000 (or 2×10^{23}).

As many readers know, the periodic table of chemistry lists the elements according to their chemical similarity. Each element is assigned a symbol and an "atomic number," Z. On this scale hydrogen (H), for example, is 1, helium (He) is 2, oxygen (O) is 8, iron (Fe) is 28, gold (Au) is 79, and uranium (U) is 92. Thus we say, "The atomic number for iron is 28," or more simply "For Fe, Z is 28."

Inside the Atom

Until the twentieth century the internal composition of atoms was not known. Only after an experiment by Rutherford in 1911 was it realized that the electrically neutral atom had a central core (nucleus) of positive charge and an outer region of negative charge. Then studies by Bohr in 1913 revealed the relationship between light and atomic structure. He assumed a motion of the electron about the nucleus similar to that of a planet around the sun.

We shall use that analogy to explain features of the atom. Picture the atom as a miniature solar system. In place of the sun there is the heavy, positively-charged central core of the atom, called the nucleus (plural: nuclei). This core is composed in general of still more basic particles—protons and neutrons. In place of the planets, there are electrons, the small, negatively-charged particles that give us electricity.

The sketch representing an atom of hydrogen shows its single electron in orbit about the nucleus, which in this case is the proton. The proton has a weight about 1800 times that of the electron. In place of gravity, the force of electrical attraction holds the particles together.

How atoms produce and absorb light also comes from Bohr's theory. Electrons have the ability to suddenly "jump" to an orbit of smaller radius, accompanied by the emission of energy in the form of

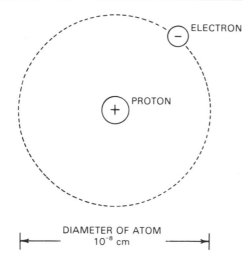

ELECTRON

PROTON

The hydrogen atom. One electron is in orbit about the nucleus (proton). The atom is a hundred millionth of a centimeter across. If the proton were the size of a golf ball, the electron would be 2000 feet away.

DIAMETER OF ATOM
10^{-8} cm

light. Similarly, an electron goes to a larger orbit when light is absorbed. If enough energy is supplied to the atom, the electron can be removed completely, leaving the positively-charged nucleus, the proton (which is also the ion, H^+).

The next most complicated chemical element is helium, used as a gas in dirigibles like the Goodyear "blimp." It has two electrons in orbit and a nucleus with two protons and two neutrons. The neutron is an electrically neutral particle weighing almost the same as the proton. Diagrams here illustrate the arrangement of particles in the helium atom and in the much more complex uranium atom.

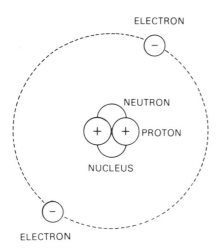

ELECTRON

NEUTRON

PROTON

NUCLEUS

ELECTRON

The helium atom. There are two electrons in orbit about the nucleus, which consists of two protons and two neutrons. The nucleus of helium is the same as the alpha particle.

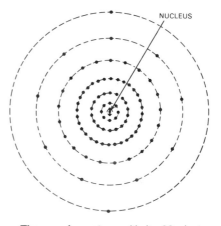

NUCLEUS

The uranium atom with its 92 electrons in orbits around the nucleus. (Courtesy of Raymond L. Murray and Grover C. Cobb, *Physics: Concepts and Consequences,* Prentice-Hall, Englewood Cliffs, New Jersey, 1970).

Radiation and Isotopes

It is important to note here that the electron is the same as the *beta (β) particle*, one form of radiation. If the two electrons are removed from the helium atom, what remains is the positively charged ion He^{++}, which is also the helium nucleus, and the *alpha (α) particle*, another form of radiation. The third main type of radiation is *light*, which takes several forms. We see by means of ordinary visible light, which comes from atoms and molecules. The x-rays used for medical examination are more energetic than ordinary light. They can come from atoms or an x-ray machine. Gamma rays have still more energy and arise in the nucleus of the atom.

Each chemical element is made up of several types of atoms called isotopes. The difference between isotopes lies in the weight of their nuclei, which is determined by the number of protons plus neutrons.

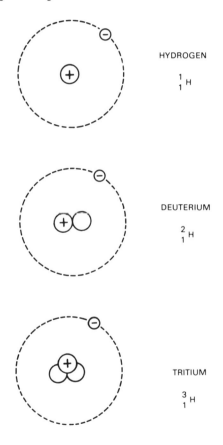

HYDROGEN

$${}_{1}^{1}H$$

DEUTERIUM

$${}_{1}^{2}H$$

TRITIUM

$${}_{1}^{3}H$$

For example, hydrogen is made up of the three species, as illustrated here: ordinary hydrogen with the proton as nucleus, deuterium with a proton plus neutron as nucleus, and tritium with a proton plus two neutrons as nucleus. We let A, called "mass number," represent the number of nucleons (protons plus neutrons) in the nucleus. Thus A = 1, 2, and 3 for the isotopes of hydrogen. Recalling that Z = 1 for hydrogen, we can write symbols to uniquely specify them: ${}_{1}^{1}H$ is ordinary hydrogen, ${}_{1}^{2}H$ is deuterium or heavy hydrogen, and ${}_{1}^{3}H$ is tritium. The superscripts are mass numbers A (nucleons); the subscripts are atomic numbers Z (protons).

The chemical hydrogen is familiar to us as one component of water H_2O. Deuterium is rare in nature, there being only one atom for every 6700 atoms of ordinary hydrogen. Tritium is a manmade isotope. These heavier species of hydrogen are the ingredients in the fusion process being developed for practical energy.

At the heavy end of the periodic table of natural elements is uranium, atomic number 92. It is composed of two main isotopes, of mass numbers 235 and 238,

The three isotopes of hydrogen. Each has one proton in the nucleus; the difference between them is only in the number of neutrons in their nuclei—none, one, and two.

with symbols $^{235}_{92}U$ and $^{238}_{92}U$. Only 0.7 percent is the lighter isotope, 99.3 percent the heavier. Each of these two isotopes plays a role in the fission nuclear reactor. A great variety of chemicals appear in the operation of nuclear processes and devices, as we shall take up later.

We have presented the abbreviated notation for isotopes because it is found in the technical literature. For most purposes, however, it is sufficient to use only the name and mass number, for example, "uranium-235" or "^{235}U."

CHAPTER 3

Radioactivity

The Process of Decay

Many of the isotopes of nature are stable, meaning that they never change. Other isotopes, both natural and manmade, are radioactive, meaning that they are unstable and can change into another form.

Radioactivity is a process in which a nucleus spontaneously disintegrates or "decays." For the simplest example, see the diagram depicting the decay of hydrogen-3 or tritium. One of the two neutrons in its nucleus changes into a proton and an electron. The new nucleus, composed of two protons and a neutron, is the same as that of an isotope of helium (helium-3). The electron emitted is called a high-speed beta particle. Radioactivity can be described by a reaction equation. That for the decay of tritium is

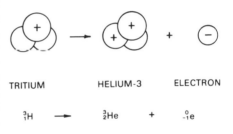

TRITIUM   HELIUM-3   ELECTRON

$$^3_1H \longrightarrow ^3_2He + ^0_{-1}e$$

The radioactive decay of the isotope tritium, shown as a diagram and by a word equation.

tritium → helium-3 + electron.

A few other important nuclear reactions that involve radioactive decay are:

uranium-238 → thorium-234 + alpha particle

cobalt-60 → nickel-60 + beta particle + gamma rays

iodine-131 → xenon-131 + beta particle.

Half-Life and Activity

One cannot predict when a particular nucleus will decay. If we could watch a group of radioactive atoms, some would decay at once, others later, still others much later. The number that decay in any second of time depends on only two things—how many there are and the nuclear species. Each isotope has its own "half-life" (t_H), which is the time it takes for half of any sample to decay. For instance, t_H for tritium is 12.3 years. Thus if we started now with 1000 atoms of tritium, after 12.3 years we would have 500 atoms, after 24.6 years, 250 atoms, and so on. The graph shows the trend with time of the number of atoms.*

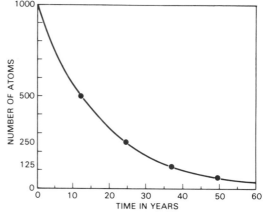

Graph of radioactive decay of tritium with time, starting with 1000 atoms.   The half-life of tritium is 12.3 years.

Disintegrations of Nuclei

The rate of decay is called the *activity*, which is the number of disintegrations per second (d/s). The activity decreases with time in the same way that the number of atoms present decreases. Thus the hazard due to the radioactive emissions decreases with time. In summary, the smaller the amount of material we have and the longer its half-life, the smaller is the activity and the safer is the sample. One rough rule of thumb when dealing with small quantities of material is that it takes 10 half-lives to eliminate a radioisotope. Since $(1/2)^{10}$ is only about 1/1000, this rule is too crude to be of value when factors of 1/1,000,000 or better are needed.

---

*It is easy to calculate the fraction f of particles that remain at any time, t. Let p stand for the number of half-lives elapsed, which is the ratio t/t_H. Then

$$f = \left(\tfrac{1}{2}\right)^p.$$

If the power p is an integer, we can do the calculation in our heads. For example, if the time is three half-lives, $t = 3t_H$, then

p = 3 and f is $\tfrac{1}{8}$.

If p is not an integer, we need to use a pocket calculator. For example, if t_H = 12.3 years as for tritium and t is 100 years, then

p = 100/12.3 = 8.13 and f is 0.00357.

Natural Radioactivity

Several of the heavier isotopes in nature are unstable and decay with the emission of an alpha particle. A good example is radium, which becomes the gaseous element radon when it decays, according to

radium-226 → radon-222 + alpha particle.

The half-life of radium is 1599 years, and the number of disintegrations per second (d/s) is around 3.7×10^{10} per gram. This value of d/s is called the curie* (abbreviation Ci) after Marie Curie, who first studied the radioactivity of uranium. Thus when we say that a cobalt-60 radiation source has an activity or "strength" of 1000 Ci, it means that the decay rate is $(1000)(3.7 \times 10^{10}) = 3.7 \times 10^{13}$ d/s. A more modern unit of activity is the becquerel (Bq), which is 1 d/s. The strength of our cobalt source is thus 3.7×10^{13} Bq.

Natural Decay Chains

The radium-radon step is but one in a long chain of radioactive processes starting with uranium-238, as shown in the following table. This radioactivity is important since it occurs in the residues called "tailings" from the mining and milling of uranium ore. The final isotope is seen to be stable lead-206. Other natural chains start with uranium-235 and thorium-232.

Some isotopes decay with the release of an alpha particle only; others yield beta particles only; but many give both beta and gamma rays. The latter are similar to visible light except that their energy is much higher and they are better able to penetrate matter.

Uranium as a mineral found in nature is only mildly radioactive, since the half-lives of both its isotopes are very long: for ^{238}U, 4.46 billion years; for ^{235}U, 704 million years.[†] However, uranium is found in many types of rock, and since several of its descendants are more radioactive than ^{238}U, many building materials give us a significant level of radiation. Radon-222, with half-life 3.82 days, is a noble (i.e., chemically inert) gas that enters buildings from the ground. It decays into radiation-emitting isotopes of polonium and bismuth, as seen in the table. Another naturally occurring radioactive element is potassium. Accompanying stable potassium-39 is weakly radioactive potassium-40 (0.017 atom%) with a half-life of 1.26 billion years.

---

*Fractions of a curie are the millicurie (10^{-3}), microcurie (10^{-6}), nanocurie (10^{-9}), and picocurie (10^{-12}). For example, one nanocurie is 37 d/s.

†From a gram of pure ^{238}U, the number of disintegrations per second is 12,400, corresponding to about 0.3 microcurie.

RADIOACTIVITY
The Chain of Natural Radioactivity Starting with Uranium and Ending with Lead
(Half-life data from "Table of the Isotopes," Norman E. Holden, in *CRC Handbook of Chemistry and Physics*, David R. Lide, Editor; branching ratios from *Table of Radioactive Isotopes*, Egardo Browne, Richard Firestone, and Virginia S. Shirley, John Wiley & Sons, 1986.)

Isotope	Name	Half-Life*	Main Radiations Emitted
$^{238}_{92}$U	Uranium	4.46×10^9y	α
$^{234}_{90}$Th	Thorium	24.1 d	β,γ
$^{234}_{91}$Pa^m 99.87% / 0.13%	Protactinium	1.17 m	β,γ
$^{234}_{91}$Pa	Protactinium	6.69 h	β,γ
$^{234}_{92}$U	Uranium	2.45×10^5y	α
$^{230}_{90}$Th	Thorium	7.54×10^4y	α
$^{226}_{88}$Ra	Radium	1.60×10^3y	α
$^{222}_{86}$Rn	Radon	3.82 d	α
$^{218}_{84}$Po 99.980% / 0.020%	Polonium	3.04 m	α
$^{214}_{82}$Pb	Lead	27 m	β,γ
$^{218}_{85}$At	Astatine	1.6 s	α
$^{214}_{83}$Bi 99.979% / 0.021%	Bismuth	19.9 m	β,γ
$^{214}_{84}$Po	Polonium	164 μs	α
$^{210}_{81}$Tl	Thallium	1.30 m	β,γ
$^{210}_{82}$Pb	Lead	22.6 y	α,β
$^{210}_{83}$Bi ~100% / 0.00013%	Bismuth	5.01 d	α
$^{210}_{84}$Po	Polonium	138 d	α
$^{206}_{81}$Tl	Thallium	4.20 m	β
$^{206}_{82}$Pb	Lead	Stable	–

* Note that numbers are rounded off to three digits (y = year, d = day, h = hour, m = minute, s = second).

Among the hundreds of radioactive isotopes (also called radioisotopes or radionuclides) are found half-lives ranging from a small fraction of a second to billions of years, as we saw in the table. A stable substance has a half-life of infinity, of course. Each isotope has its own half-life, which is unaffected by any chemical treatment.

We have emphasized the "natural" radioisotopes, i.e., those found in nature. "Artificial" or manmade radioisotopes can be produced by irradiating (bombarding) stable nuclei by various particles such as protons, neutrons, alpha particles, and deuterons. Radioisotopes also are by-products of nuclear fission, as will be seen later. The figure below, showing some nuclear reactions, provides information on neutron absorption and radioactivity, for use in later sections.

NEUTRON ABSORPTION

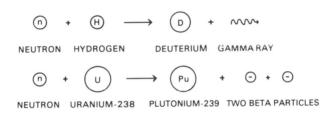

RADIOACTIVE DECAY PROCESS

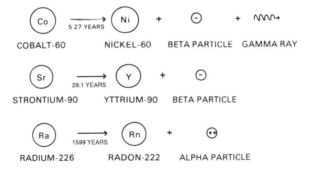

Some nuclear reactions. The symbol D is often used for deuterium (heavy hydrogen $^{2}_{1}H$); the wavy line with arrow is intended to suggest that the gamma ray is an electromagnetic wave; half-lives of the decay processes are listed.

CHAPTER 4

Kinds of Radiation

Particle Energy

We introduced three important types of radiation (α, β and γ) in the preceding section. Recall that the beta particle is the same as the electron, which is a negatively charged particle with very small mass, found in atoms. The beta particle is emitted from the nucleus in radioactive decay also. An example is the reaction

strontium-90 → yttrium-90 + electron.

Strontium-90 is an important nuclear waste because of its relatively long half-life of 29.1 years.

The energy of a beta particle depends on the process from which the particle comes. Also, its energy determines its ability to penetrate matter and cause radiation damage.

The typical unit in which particle energies are expressed is the "electron volt." We can understand the unit by doing an experiment. Picture a one-volt battery connected to two plates of opposite polarity. If an electron is carried across the gap, the energy given to the electron is said to be one *electron volt*, abbreviated *eV*. Now suppose that one electrically accelerates a beta particle to the maximum energy it has on emission from strontium. This time the energy is much larger—546,000 eV or 0.546 MeV (million electron volts). These strontium-90 beta particles are still less energetic than some, e.g., those of about 4 MeV from gallium-66 and silver-112.

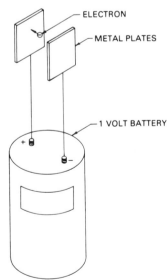

The electron volt as a unit of energy. One kilowatt-hour of energy, the amount used by a 1000-watt appliance in an hour, is equal to 2.25×10^{25} electron volts.

Remember that the alpha particle is the helium nucleus. It is emitted by reactions such as the decay of uranium-238 into thorium-234. The alpha particles have an energy of 4.2 MeV, typical for alphas emitted by heavy natural radioactive isotopes. The neutron is produced by certain reactions involving the bombardment of a nucleus of an element such as beryllium by alpha particles from radium decay. The final nuclear reaction is

beryllium-9 + helium-4 → carbon-14 + neutron.

Such processes gave rise to the neutrons that were used in the discovery of fission. However, the neutron is not involved in the radioactivity of nuclear wastes since practically no nuclei decay with neutron emission.

On the other hand, neutrons from outer space continuously produce radioactive carbon by the reaction

neutron + nitrogen-14 → carbon-14 + proton.

The radioactive carbon-14 is used to find the age of archeological items, as noted in Chapter 10. Neutrons can also produce new elements. Neutrons absorbed in uranium-238 give rise to the artificial (manmade) isotope plutonium-239. Plutonium-239 is an alpha particle emitter that can serve both as a nuclear reactor fuel or as a weapon, as we shall see later.

Gamma Rays

The gamma ray can be imagined to be a burst of energy, a particle, or a wave. Each view is correct in some sense. Gamma rays are at the high-energy end of what is called the electromagnetic spectrum. As shown in the chart here, it includes radio, microwaves, infrared, visible light, ultraviolet, x-rays, and gamma rays, in increasing order of energy.

For nuclear purposes, though, it is often better to think of a gamma ray as a particle of light or a bundle of energy called a photon. Many radioisotopes are gamma emitters. For example, cobalt-60 gives two gamma rays of energy around 1.25 MeV that can be used in medicine in either diagnosis or cancer therapy, as an alternative to x-rays. Gamma rays are emitted at the same time as beta particles from many of the radioactive by-products of the fission process. Such substances are part of the radioactive waste as discussed in Chapter 7.

Radiation Spreading and Stopping

Let us now study the behavior of these various particles in air or in a vacuum. If we were in a laboratory with a small piece of radioactive material, say strontium or cobalt, the particles would come out in all directions. As seen in the sketch, radiation from such a source is

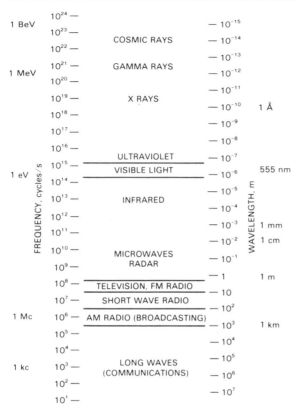

The electromagnetic spectrum, showing that radio-TV, visible light, and nuclear radiation are all waves.

Comparison of the spreading of two forms of radiation— visible light and nuclear radiation. If the distance from the source is doubled, the intensity is divided by four.

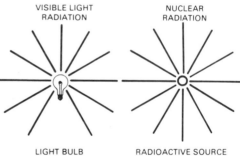

similar to that from a light bulb. First, the rays become less intense with distance because of "inverse square spreading."* Prolonged exposure to radiation of either kind can be harmful. Infrared rays can burn the skin; the ultraviolet light component can damage eyes and cause skin cancer. Alpha particles, beta particles or gamma rays can damage body tissue as well.

---

*The intensity varies inversely as the square of the distance because all radiation goes through spheres of area $4\pi r^2$.

We can protect ourselves by backing away from the sources. Or, as shown in the sketch, we can interpose some solid material between us and the source of radiation. Each particle is slowed or stopped by collisions with atoms of the substance—in some cases by interactions with the electrons, and in others with the nuclei. A sheet of paper would be enough to stop alphas, but it would take a sheet of aluminum metal 1/25 inch thick to stop betas, such as those from strontium-90. These particles are thus said to have a certain "range,"

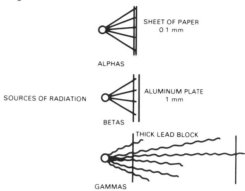

Stopping of radiation by various shields.

which increases with particle energy, of course.

The intensity of gamma rays is merely reduced by passage through matter, just as light intensity is reduced by fog. A half thickness "t_H" can be defined, reminiscent of half-life. It is the distance it takes to cut the intensity of gamma radiation in half. For the cobalt gammas, the element lead has a t_H of close to one centimeter (cm), making it a useful shield for medical radiation diagnosis or treatment.

How radiation interacts with matter on a submicroscopic level is well understood. We can visualize, for example, the effects of a collision between a gamma ray and a simple atom, as sketched here. There are three possible events: (a) scattering, in which the gamma ray bounces off the electron; (b) ionization, in which the energy of the gamma ray goes into removing the electron from the atom, leaving an ion; and (c) pair production, in which the gamma ray energy is converted into the mass of two particles—an electron and a positron (positively charged electron). This process, pair production, illustrates Einstein's theory that energy and mass are two forms of the same thing.

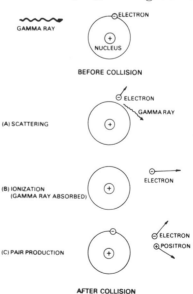

Effects of gamma radiation. There are three possibilities—scattering, ionization, and electron-positron pair production.

CHAPTER 5

Biological Effects of Radiation

Radiation Takes Many Forms

Modern people are already familiar with quite a few forms of radiation. They make use of visible light, both natural and artificial, in all their activities. They enjoy microwave ovens, radio, and television, which operate on low-frequency electromagnetic waves. They have experienced the well-known form of radiation exposure, sunburn.

Less familiar but real is the unfelt but continued bombardment by radiation from space and the earth. Also little understood are the radiations from nuclear devices and products. In this chapter we shall study the effects of rays and particles from radioactive materials, with special attention to low-level radiation.

Radiation and Living Cells

When high-speed particles such as alphas, betas, gammas, or neutrons strike living tissue, they slow down and stop just as if they had hit paper or aluminum or lead. The energy of motion of the individual particles is imparted to the biological cells as localized heat. Molecules of the cells are changed, or atoms are converted into ions, or the atomic nuclei are displaced from their positions.

The disruptive action caused by absorbed energy impairs human body cell functions. If the amount of radiation received is extremely small, there may be no significant damage. If the amount is very large, radiation sickness, genetic effects, or death may result. The words "large" and "small" are, of course, relative, so we need to express the amounts of radiation received in terms of numbers.

The biological effect of radiation, which we call dose or dosage, depends on the amount of energy absorbed and also on the type of radiation. Each radiation has a different effect on tissue. For example, neutrons are about ten times as damaging as x-rays, gamma rays, or low-energy beta particles. Alpha particles are about 20 times as damaging. These factors that represent biological effectiveness are taken into account when the dosage is expressed in *rems*.* The rem

---

*The actual energy absorption is measured in rads, with 1 rad as 0.01 watt-second energy absorbed per kilogram of tissue weight. The factors of biological effectiveness are applied to convert the number of rads into the number of rems. Modern scientific units are the gray (Gy), which is 100 rads, and the sievert (Sv), which is 100 rems.

is a unit of dosage just as the inch is a unit of length. For low radiation levels encountered regularly by human beings we use the millirem (mrem) as 1/1000 rem.

Mankind throughout all existence has been exposed to radiation from natural sources, including the ground, radionuclides in the body, and from space in the form of cosmic rays. Those rays bombard our atmosphere and all beings on the earth's surface. Another source that has only recently been recognized as significant is the element radon. As noted in Chapter 3, radon-222 is the result of decay of radium-226 and is the parent of some radioactive bismuth and polonium isotopes. As a colorless, odorless, and invisible gas, radon seeps into buildings from the ground and is breathed by the occupants.

The diagram shows the average individual American's annual exposure, which is estimated to be 360 mrems. Doses and percents from the various sources are shown. The dose varies from one state to another because of differences in the uranium content of the earth. The dose varies with altitude, which affects the amount of screening effect of the atmosphere. The radioactive isotope potassium-40, as a component of the trace element potassium in foods, becomes a part of our bodies.

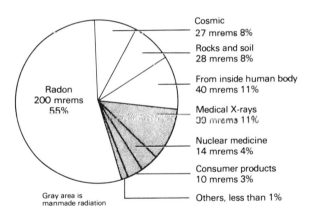

Sources of radiation exposure. The estimated annual dose to an average American is 360 millirems, according to the National Council on Radiation Protection and Measurements Report No. 93, 1988.

Radon 200 mrems 55%

Cosmic 27 mrems 8%

Rocks and soil 28 mrems 8%

From inside human body 40 mrems 11%

Medical X-rays 39 mrems 11%

Nuclear medicine 14 mrems 4%

Consumer products 10 mrems 3%

Others, less than 1%

Gray area is manmade radiation

Hazards to the Body

Questions often asked when there is extra radiation present are "How much danger is there?" and "What effect will the radiation have?"

There are two general classes of radiation effects—somatic, meaning damage to body tissue, and genetic, referring to hereditary characteristics. The main somatic effect is cancer such as leukemia or tumors. Genetic effects involve transmitted abnormalities or early death.

Each tissue and organ of the body has a different degree of sensitivity to radiation effects. The blood-forming tissue, the gastrointestinal tract, and the gonads are some of the more readily affected. The body's external layer of skin provides some protection from bombardment by alpha particles because the particles penetrate only a short distance. However, radioactive material that emits alpha particles can be very hazardous if taken into the body.

Evidence of these physiological effects of massive radiation dosage comes from many sources: laboratory experiments on lower animals such as mice; observation of side effects of radiation treatment of certain diseases; data from the past on radium poisoning of people who painted luminous watch dials; the incidence of lung cancers among uranium miners who worked with inadequate ventilation of radioactive radon gas; and, finally, studies on the survivors of the atomic bombing of Hiroshima and Nagasaki.

Studies of radiation effects lead to two conclusions. The first is that there is no direct evidence of genetic damage in humans, even though mutations have been observed in plants and lower animals. The second is that a single radiation dose of around 400 rems will be fatal to half of those who receive it, while half will survive, perhaps with some impairment of function. Since such large doses are rare, we are more interested in the effects of small doses.

Low-Level Radiation and the Linear Model

The subject of low-level radiation is subtle and complex. Let us start with some facts. First, there are no directly observable effects on human beings of a radiation dose smaller than 10 rems. Second, people are exposed continuously to a radiation background from cosmic rays and the ground of about 300 mrems per year in the U.S. and several times larger in some countries such as India or Brazil, where there are large natural deposits of radioactive minerals. Third, there is no indication of a geographic difference in health attributable to radiation. It is logical to assume that there is some level of dosage below which there is no permanent effect because of the body's ability to recover. On the other hand, it can be argued that any amount of radiation is harmful.

We are faced with a large gap in information between the extremes of zero radiation (and zero effect) and very high radiation doses. The simplest assumption is that radiation effect is directly proportional to radiation dose. Thus, as shown in the drawing, one starts with the available data on effects of high radiation doses and draws a straight-line graph down to zero. This solid line expresses an important statement, often called the "linear model" or "linear hypothesis." The other two dashed curves are possible alternatives. The lower one involves a threshold dose, below which nothing happens. It is safer to use the straight-line graph since it predicts a larger effect for a certain dose.

The linear graph is thus said to be "conservative." On the other hand, few believe that the upper dashed curve is correct.

Scientists currently represent hazard versus dose by an improved formula that predicts the effect to vary directly with the dose for low values and to vary as the square of the dose for high values. This "linear-quadratic" model is reasonable if damage to sensitive tissues at low doses is by single radiation "hits" while that at high doses is by two or more "hits." For further information, the first chapter of BEIR V* is highly recommended.

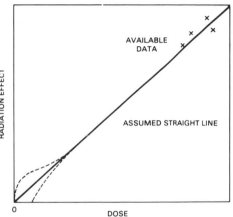

The effect of radiation dose. The solid line is assumed as an extension of data on fatalities from high radiation doses. The lower dashed curve is more likely to be correct than the upper dashed curve.

How Sure Can We Be?

Since we cannot observe the effect of low-level radiation on an individual, we must view the effect statistically. Thus if a large number of people all receive the same small dosage, it is presumed that there will be a few fatalities over the course of years. A unit of total population dose is called the person-rem. For instance, if each in a group of 100 people received 5 mrems (0.005 rem), the population dose would be 100 x 0.005 = 0.5 person-rem. The linear model yields a prediction of one additional cancer death for each 2500 person-rems.

We can apply this to the Three Mile Island reactor accident of March 1979 near Harrisburg, Pennsylvania. It was estimated that the 2 million people living within a 50-mile radius received a total of 3300 person-rems. The predicted additional cancer deaths (over and above those from other causes) would thus be 3300/2500 = 1.32. Assuming that there would be an equal number of radiation-caused genetic deaths, the total comes out fewer than three additional fatalities attributable to the accident. Looked at in another way, each of the two

---

*Health Effects of Exposure to Low Levels of Ionizing Radiation BEIR V, National Academy Press, Washington, DC, 1990.

million people in the Harrisburg area could say that his or her increased chance of dying as a consequence of the incident was about one in 750,000.

The Chernobyl reactor accident of 1986 resulted in the exposure of a few people to high radiation doses, well above the lethal level. It also gave millions of people in Europe small increases over natural background. Based on statistics, one would predict that thousands would eventually die prematurely.

The effects of various radiation doses over a wide range of values are shown in the diagram. The scale goes down by ten at each mark. For example, the dose at a nuclear plant boundary is lower by a factor of 100,000 from a fatal figure.

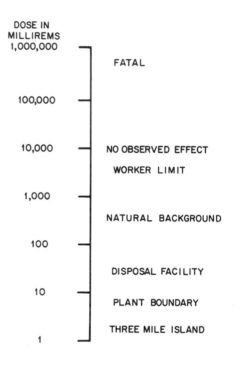

What Is Risk?

We can understand risk better by examining data on fatalities from various causes, including diseases and accidents. The following table gives U.S. figures for the year 1989. Let us use that table to estimate the chance of death each year from some common causes. The number of fatalities from heart disease was 720,862 out of a total population of around 252,180,000. The chance of dying of heart disease each year is thus 1 in 350. From the figure on motor

Examples of radiation doses. Note that the scale is logarithmic.

vehicle accidents, 43,536, the chance is 1 in 5792. Note that these calculations are averages over the whole population.

Clearly, one's chances of being killed on the highway greatly exceed those of being killed by radiation; but most people feel that they have control of their fate when they drive but do not if a nuclear reactor is operated by someone else. Also, automobile accidents are common, easily understood, and only a few people are affected in each accident, while reactor accidents are rare and mysterious, and it is possible that a large number of people could be affected. We know that many people deliberately take chances with their lives by smoking, drinking, and taking drugs. Others find excitement in hazardous sports. Also the perception of risk varies widely among different groups of people, as noted by Upton in a readable and informative article.*

The question is often asked "Why do we not know the effect of low-level radiation more precisely?" The basic reason is that there are too many competing causes of injury. Radiation can cause cancer, but so can exposure to many kinds of foods, drugs, chemicals, and other pollutants. To demonstrate this statistical situation, suppose that we made a very careful study of 1000 atomic workers over their lifetimes and found

Major Causes of Death in the U.S. for 1991

(From Table 27, Monthly Vital Statistics Report, August 31, 1993, U.S. Department of Health and Human Services.)

The total population was approximately 252,180,000.

Heart disease		720,862
Cancer		514,657
Stroke		143,481
Lung disease		90,650
Accidents, total		89,347
motor vehicle	43,536	
other	45,811	
Pneumonia, influenza		77,860
Diabetes		48,951
Suicide		30,810
HIV infection		29,555
Homicide		26,513
Liver disease		25,429
Kidney disease		21,360
Blood poisoning		19,691
Hardening of arteries		17,420
Infants at birth		16,781
All other causes		296,151
Total		2,169,518

that 241 of them died of cancer. The table shows that the number of such fatalities in an average group of 1000 deaths in the population would be 237. It would appear that four excess deaths are due to radiation exposure. However, statistical variations of at least that number are expected among samplings of people. For example, there is a difference of more than 6 cancer deaths per 1000 between 1989 and 1991, using the total of more than 2 million fatalities. In any actual sample of 1000 deaths, the fluctuations will be even greater. It is not possible to conclude that radiation caused any of, all of, or even more than the four extra fatalities. Another factor that clouds the issue is the "healthy worker effect," in which people in an industrial group studied tend to take better care of their health than average citizens. Although continued studies should and will be made, there is not likely to be a major improvement in accuracy of the estimated effect of low-level radiation.

---

*Arthur C. Upton, "The Biological Effects of Low-Level Ionizing Radiation," *Scientific American*, February 1982, p. 41ff.

Comparison with Chemical Wastes

Radioactive wastes from various nuclear processes, one source of harmful radiation, should be viewed in perspective. Certainly the hazard seems very ominous because some of the isotopes last for thousands or even millions of years. Some chemical poisons, however, such as arsenic, lead, and mercury last forever. In a sense, radioactive materials are degradable, while the poisons are not.

As pointed out by Cohen,* the "potential" hazard is greater for several toxic chemicals than for radioactive wastes. He states that the lethal doses produced per year in the U.S. of chlorine, phosgene, ammonia, cyanide, and barium all exceed those of nuclear waste. He also notes that the chemicals are more accessible than are radioactive wastes.

In recent years, the public has a heightened awareness and concern about risks, as the result of new information on chemical contamination at places such as Love Canal, and on the effects of pesticides on animal life. However, a number of substances have been found to be far less hazardous than portrayed in the press. Examples are saccharin, alar on apples, dioxin, and asbestos.

*Bernard L. Cohen, "High Level Radioactive Waste From Light-Water Reactors," *Reviews of Modern Physics*, January 1977.

CHAPTER 6

Radiation Standards and Protection

When high-energy atomic and nuclear radiation was first discovered early in this century, many experimenters received excessive doses of radiation. The discovery of x-rays led to widespread use of this radiation for medical diagnosis. Scientists and doctors were not aware of the biological damage that such radiation could produce and of the need for great caution in administering x-rays. As a result there were a number of radiation burns and fatalities. In painting the naturally radioactive element radium on watch dials to make them luminous in the dark, workers who pointed their brushes with their lips received excessive radiation and many died of leukemia.

Standards for Protection

Over the years, however, more information became available and greater safety measures were developed. Certain official organizations were established and standards for maximum allowed exposure were set. The limits on dosage have generally come down. New sciences appeared: radiology, health physics, and radiation protection. In the early days, the test of radiation dose was reddening of the skin. Such primitive methods have been replaced by detection using sensitive electronic instruments and photographic film. There are well-known rules for ensuring the safety of patients who receive x-rays. We now have detailed tables of standard values of the limits on concentrations of radionuclides in air or water and information on the allowed dose to specific tissues and organs of the body. Organizations responsible for recommending new standards are the International Commission on Radiological Protection (ICRP), formed in 1928, followed the next year by the United States counterpart, the National Council of Radiation Protection and Measurements (NCRP).* The NCRP has specified dose limits for nuclear plant workers.

---

*The NCRP consists of 75 council members from universities, hospitals, national laboratories, and government agencies. The body is not an official government agency, however.

What Determines Radiation Dosage?

The radiation dose that we could receive depends in part on the amount of radioactivity in the air that we breathe or the water that we drink. Limits on concentration of a radionuclide are obtained by use of the science of health physics. Several questions are considered:

- First, what is the affinity of the isotope for an organ of the body? It may be an element that deposits in the bone, such as strontium-90; it may deposit mainly in the thyroid gland, such as iodine-131 or iodine-129; it may be a gas that affects the lungs, such as radon-222 or krypton-85.

- Second, what radiations does it emit? If it gives off only soft (low-energy) betas, as carbon-14 does, the hazard is less. But if it emits alphas, as plutonium does, or hard (high-energy) gammas, as cobalt does, the hazard is greater.

- Third, what is its half-life? If it is extremely short, the isotope may be practically gone before the body takes it in. If it is very long, the activity is low and thus little radiation is received.

- Finally, how does the body react when the isotope is ingested? Certain heavy-element oxides, if taken orally, would not remain in the body, but would soon be eliminated. If breathed as large particles, however, they might lodge in the lungs and stay for a long time. An isotope of ordinary body chemical such as hydrogen, carbon, oxygen, nitrogen, sodium, or chlorine, would become a part of the body, but would also be eliminated rather soon because of normal body processes. Thus hydrogen-3 (tritium) and carbon-14 are readily eliminated as water and carbon dioxide.

The radiation effect of a radioisotope that has entered the body thus depends on how rapidly the substance is removed by a combination of radioactive decay and biological elimination. Each process has a half-life. A formula involving the half-lives of the two processes is used to find the "effective" half-life of the substance.* For example, the ordinary half-life for tritium is 12.3 years and its biological half-life is 12 days, but its effective half-life is only 11.97 days.

---

*If t_H is the radioactive half-life and t_B is the biological half-life, the effective half-life t_E is found from the formula

$$1/t_E = 1/t_H + 1/t_B .$$

Values of t_B are: cesium-137, 70 days; strontium-90, 50 years; and plutonium-239, 200 years, as listed in ICRP Publication 2.

Radiation Protection Practices

A special document known as the BEIR report* provides some general guidance on radiation protection standards. The principles in simplified form are as follows:

1.   Allow no exposure unless there is an important benefit.
2.   Protect the public but do not waste large amounts of money on small improvements.
3.   Radiation risks should be small compared with normally accepted risks.
4.   The average dose to many persons should be much less than that for an individual.
5.   Medical radiation exposure can and should be reduced by avoiding mass x-rays, by inspection and licensing, and proper training and certification.
6.   Cost-benefit analysis should be applied to the nuclear industry.
7.   Extraordinary efforts should be made to minimize the risk of a serious reactor accident.
8.   Occupational and emergency exposure limits should be set.
9.   Studies should be made on the relationship of radiation and ecology.
10.   Good estimates and predictions should be sought.

Regulation of Radiation

For many years prior to 1994, regulatory limits on radioactivity in air and water were expressed as maximum permissible concentrations (MPCs). As of 1994, however, the Nuclear Regulatory Commission (NRC) required the use of two new measures. The first of these is the "annual limit of intake," abbreviated ALI. For example, for a nuclear plant worker, the ALI for oral ingestion of cesium-137 is 100 µCi and that for inhalation is 200 µCi. The second measure is the "derived air concentration" (DAC), being the activity in air that would lead to an ALI if breathed continuously during work hours. For ^{137}Cs, this figure is 6×10^{-8} µCi/ml.

---

*The Effects on Population of Exposure to Low Levels of Ionizing Radiation, Report of the Advisory Committee on the Biological Effects of Ionizing Radiation, National Academy of Science, National Research Council, November 1972. This document is called BEIR I. Later reports by the committee are BEIR II (1977), BEIR III (1980), BEIR IV (1988), and BEIR V (1990).

Values of the ALI and DAC as they vary with radionuclide are tabulated in the publication, *Code of Federal Regulations 10 Energy,** Part 20, Standards for Protection Against Radiation, abbreviated as 10 CFR 20. Limits are also given on the concentration of radionuclides above natural background in effluents of air and water from a nuclear facility. The table shows the maximum concentrations for some commonly found radionuclides. These figures are lower than those allowed for plant workers. For example, the air limit for ^{137}Cs is 2×10^{-10} µCi/ml, which is 1/300 of the occupational number. Note that the concentration limits for weak beta particle emitters ^{3}H (tritium) and ^{14}C are much higher than those for energetic alpha-emitting neptunium and plutonium. It turns out that the chemical toxicity of uranium can exceed the radioactive hazard. If several isotopes are in the air or water, their radiation effects must be added.

Maximum Effluent Concentrations, µCi/ml, Above Background for Selected Radionuclides
(From: *Code of Federal Regulations 10 Energy,* January 1, 1994, Superintendent of Documents)

Isotope	Air	Water
Carbon-14	3×10^{-9}	3×10^{-5}
Cesium-137	2×10^{-10}	1×10^{-6}
Hydrogen-3	1×10^{-7}	1×10^{-3}
Iodine-129	4×10^{-11}	2×10^{-7}
Iodine-131	2×10^{-10}	1×10^{-6}
Krypton-85	7×10^{-7}	—
Neptunium-237	1×10^{-14}	2×10^{-8}
Plutonium-238	2×10^{-14}	2×10^{-8}
Plutonium-239	2×10^{-14}	2×10^{-8}
Radium-226	9×10^{-13}	6×10^{-8}
Radon-222	1×10^{-10}	—
Strontium-90	3×10^{-11}	5×10^{-7}
Uranium-235	3×10^{-12}	3×10^{-7}
Uranium-238	3×10^{-12}	3×10^{-7}

The NRC specifies in 10 CFR 20 the maximum annual occupational dose, taking account of the sensitivity to radiation of body organs and tissues. As described in the regulation,[†] organ weighting factors are used to arrive at a "total committed effective dose equivalent," which considers both external and internal exposure. The basic limit is 5 rems per year, with somewhat higher figures allowed for the lens of the eye (15 rems/yr) and the skin, hands, and feet (50 rems/yr).

For the general public, the limits are considerably lower, being only 0.1 rem (100 millirems) per year. For members of the public living near a radioactive waste disposal site, the dose is limited still further, to 25 millirems per year. The reason for the difference between limits for workers and the public is that workers voluntarily accept employment where radiation may be found. On the other hand, they have

---

*This book, reissued yearly by the Superintendent of Documents, includes regulations of the Nuclear Regulatory Commission (NRC), based on specifications by the Environmental Protection Agency (EPA).
†10 CFR 20, Section 20.1003, Definitions.

protection through monitoring and control practices such that their exposure seldom reaches the regulatory limits. There are special rules that further limit the dose for minors and pregnant women.

"As Low As Is Reasonably Achievable"

The NRC's requirements on a licensee's radiation protection program as in 10 CFR 20.1101 and 10 CFR 20.1003 are "...shall use to the extent practicable, procedures and engineering controls based on sound radiation protection principles to achieve occupational doses and doses to members of the public that are as low as is reasonably achievable (ALARA)." The acronym "...means making every reasonable effort to maintain exposures to radiation as far below the dose limits...as is practical...taking into account the state of technology, the economics of improvements in relation to the state of technology...(and) to benefits to the public health and safety, and other societal and socioeconomic considerations, and in relation to utilization of nuclear energy and licensed materials in the public interest." The application of this principle led the NRC (in 10 CFR 50 Appendix I) to limit annual whole-body dose to any individual outside a nuclear power plant to 3 millirems via liquid and 5 millirems via air. The annual total release of radioactive material (excluding tritium and dissolved gases) should be no more than 5 curies.

The use of ALARA is seen to involve judgment based on information on both radiation and cost. The logic of the principle has three parts. First the regulatory dose limits must be met. Then efforts should be made to do even better. The NRC suggests the use of $1000 per person-rem in cost-benefit calculations (10 CFR 50 Appendix I). Finally, zero risk is not sought since it would obviously entail infinite cost.

Nuclear plants meet the standards set by the Nuclear Regulatory Commission by using a combination of sensitive radiation-detection equipment and certain procedures. The photograph shows a plant worker using a portable radiation instrument. Radiation levels are measured by radiation detectors and dosimeters, while radioactive contamination is determined by samples of water or air that might contain radioactive isotopes. For improved accuracy of measurement, large water samples can be evaporated to increase the radioactive concentration, and large air samples can be passed through filters that collect the contained radioisotopes.

The values of the maximum concentrations in the foregoing table are extremely small, but the very fact that the isotopes are radioactive and emit radiation makes it possible for them to be detected. Even though radiation cannot be seen, felt, heard, smelled, or tasted, its presence can be sensed by radiation detectors and the amount of hazard can be measured.

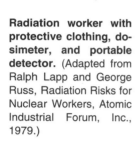

Radiation worker with protective clothing, dosimeter, and portable detector. (Adapted from Ralph Lapp and George Russ, Radiation Risks for Nuclear Workers, Atomic Industrial Forum, Inc., 1979.)

How To Protect Against Harm

Radioactive particles can enter the body by ingestion—by eating, drinking or breathing. To protect against radiation hazard, it is necessary to isolate the source of radioactivity or to render it harmless. Storing the material in a safe place until it decays solves the problem if the half-life is short, for example, iodine-131 with its half-life of 8 days. Releasing the material into very large volumes of air or water can dilute the activity to below the maximum concentrations shown in the preceding table. Mixture with dirt in the ground is possible in some applications.

To provide protection against radiation that is external to the body, three factors can be used—distance, time, and shielding. A person is safer the farther from the source of radiation, the shorter the time of exposure, and the thicker the shielding. The following sketch shows some ways dosage is minimized. The radiation warning symbol is used universally; rope barriers remind workers of potential hazard.

Exposure to radiation from wastes is prevented by use of protective containers and by isolation of the radioactive material, as discussed later.

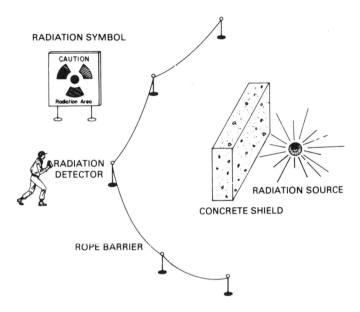

RADIATION SYMBOL

CAUTION

Radiation Area

RADIATION DETECTOR

RADIATION SOURCE

CONCRETE SHIELD

ROPE BARRIER

Examples of methods of protection against radiation.

CHAPTER 7

Fission and Fission Products

The Splitting of the Nucleus

The radioactive wastes that are by-products of nuclear power generation arise mainly from the fission process. Fission is the splitting of a nucleus into two parts, triggered by absorption of a neutron.*

Using the sketch, picture the stages, beginning with a neutron approaching the nucleus of uranium-235. The neutron is absorbed and forms uranium-236, a compound nucleus. This particle is unstable, with some forces trying to hold it together, others trying to disrupt it. Imagine internal vibrations being set up that cause the shape of the nucleus to be distorted. Electric repulsion forces then cause the nucleus to separate into two fragments that fly apart at high speed. These fission fragments are the radioactive "fission products." Their energy of motion will eventually be converted into useful heat. As shown, several

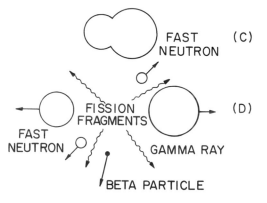

Stages in the fission process. A slow neutron approaches the uranium-235 nucleus in step (A), is absorbed to form uranium-236, as in (B). The nucleus becomes distorted as in (C), and the two fragments fly apart at high speed and various radiations are emitted, including neutrons, as in (D). (Adapted from Raymond L. Murray, *Nuclear Energy,* Pergamon Press, 1993.)

---

*Fission happens spontaneously (i.e., the isotope does not need to absorb a neutron) in only a few isotopes such as californium-252.

neutrons, gamma rays and beta particles are released during this violent separation.

Nuclei of uranium-236 may split in many different ways. If three neutrons were released, the most likely atomic weights would be around 140 and 93. The graph shows which fission product masses are more likely. It tells us, for example, that isotopes with A = 134 are produced in about 7 percent of the fissions. We see that masses 90 and 137 are quite abundant.

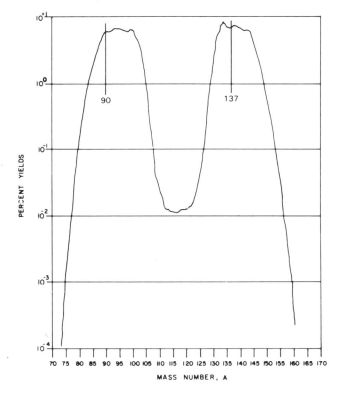

Yield of fission products according to mass number. Many elements and isotopes are present, both stable and radioactive.

Typical nuclear reactions are:

uranium-235 + neutron → uranium-236
uranium-236 → krypton-90 + barium-144 + 2 neutrons.

The uranium-236 splits into fission fragments in about 85 percent of the cases but merely releases a gamma ray in 15 percent. Note that although we talk about "^{235}U fission" it is really the ^{236}U that divides.

The terms "fissionable" and "fissile" are often used to describe types of nuclei. All nuclei can be made to undergo fission if the bombarding neutron has high enough energy. In other words, all nuclei are fissionable, but only a few, the fissile isotopes, can be fissioned with slow (thermal) neutrons. These are uranium-235, uranium-233, plutonium-239, and plutonium-241. Another technical term, "fertile," refers to an isotope that can be converted into a fissile one by absorbing a neutron. Examples are uranium-238, thorium-232, and plutonium-240.

Several neutrons come from the fission process. Thus a chain reaction is possible, in which neutrons cause fissions that release neutrons that produce more fissions and so on. When conditions are right, we have a steadily operating source of energy, a "critical" nuclear reactor. The minimum ingredients are uranium and neutrons; many other components are needed to take away heat, provide structural strength, and permit control and safety. Uranium, through its fissile isotope uranium-235, is the fuel for a nuclear reactor; it is "burned" in the sense that the absorption of a neutron causes energy to be released.

Radioactive Fission Products

The fission fragments are individual nuclei. Collectively they are fission products, about 800 different isotopic species. Most of them are radioactive, emitting a series of beta particles and gamma rays, with half-lives that generally start out short and become longer in later stages of the chain.

Two important chains are depicted here, with the half-lives for each decay shown. Note that as beta particles are released, the atomic number of the isotope increases. These chains lead to the nuclear waste isotopes strontium-90, half-life 29.1 years, and cesium-137, half-life 30.3 years. Some other important fission products and their half-lives are: iodine-131, 8.04 days; cerium-144, 284.6 days; cesium-134, 2.065 years; krypton-85, 10.73 years; technetium-99, 213,000 years; and iodine-129, 17 million years. Other isotopes produced are hydrogen-3 (tritium), 12.32 years, and carbon-14, 5715 years. Several fission product isotopes are valuable for use in research, medicine,

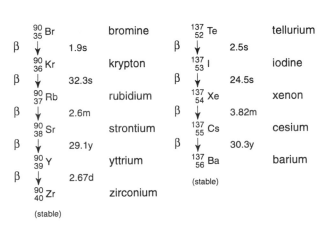

Two fission product decay chains leading to the important waste isotopes strontium-90 and cesium-137. These isotopes dominate the radioactivity in the waste for several hundred years. See also the following graph. (Data from *CRC Handbook of Chemistry and Physics*.)

$^{90}_{35}$Br		bromine
β ↓	1.9s	
$^{90}_{36}$Kr		krypton
β ↓	32.3s	
$^{90}_{37}$Rb		rubidium
β ↓	2.6m	
$^{90}_{38}$Sr		strontium
β ↓	29.1y	
$^{90}_{39}$Y		yttrium
β ↓	2.67d	
$^{90}_{40}$Zr		zirconium
(stable)		

$^{137}_{52}$Te		tellurium
β ↓	2.5s	
$^{137}_{53}$I		iodine
β ↓	24.5s	
$^{137}_{54}$Xe		xenon
β ↓	3.82m	
$^{137}_{55}$Cs		cesium
β ↓	30.3y	
$^{137}_{56}$Ba		barium
(stable)		

and industry. New commercial applications such as food preservation by radiation treatment may increase the demand for such radionuclides.

When fuel is first installed in a nuclear reactor, it has a rather low radioactivity, mainly from uranium-235 and uranium-238. Upon irradiation by neutrons, fission products are produced and decay, but they build up to a rather constant inventory. The decay with emission of betas and gammas contributes to the useful heat energy of the reactor, but also lasts for years after fuel is removed from the reactor. The graph shows the slow decline in rate of heat generation.

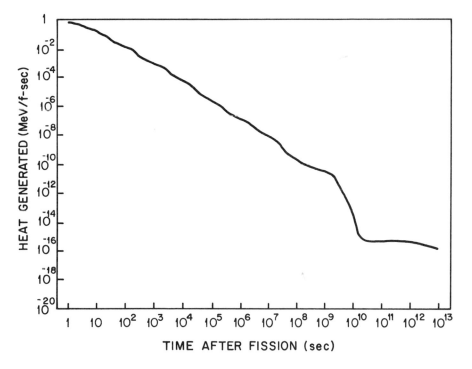

Heat generated by fission product isotopes as it depends on time after removal from a nuclear reactor. (After Bernard L. Cohen, *Scientific American,* June 1977.)

Energy from Fission

Each fission reaction releases a very large amount of energy, about 200 MeV, i.e., 200,000,000 eV. The fission fragments carry about 166 MeV of this total, and the neutrons 5 MeV, and about 20 MeV goes to other radiations. The 200 MeV figure is to be contrasted with the low energy obtained by burning a chemical fuel. When hydrogen atoms react with an oxygen atom to produce a molecule of water, the energy yield is only 3 eV.

Per pound of fuel burned, the energy from nuclear fission is millions of times that from burning a chemical fuel such as coal or oil. Hence, the weight of nuclear wastes per kilowatt-hour of energy produced is extremely small. One gram of waste results from one megawatt-day of reactor heat energy production.* In contrast, there is a weight of some 2.5 tons of waste solids and gases from burning of a fossil fuel such as coal to produce the same amount of heat.

Fission is sometimes confused with another important nuclear process, fusion. The latter brings together two light isotopes such as deuterium and tritium to "fuse" or combine and release large amounts of energy. The fusion process, which is the source of the sun's energy, is being investigated as a potential source of energy. If successful technically, it would produce relatively few radioactive wastes. Few predict, however, that fusion will contribute to the world's energy in the twentieth century.

---

*Recall that 1000 watts is a kilowatt and a million watts is a megawatt.

CHAPTER 8

The Manhattan Project

Wartime Wastes

Large volumes of nuclear wastes were produced in World War II in the effort to collect the plutonium needed for atomic bombs. This work was part of a defense effort with the code name Manhattan Project. A review of its history* will reveal how the present waste situation arose.

We recall that direct involvement of the United States in the war began in December 1941 following the attack on Pearl Harbor by the Japanese. The Allies—consisting mainly of the United States, Great Britain, Russia, Canada, and France—were opposed by the Axis powers—Germany, Japan, and Italy.

The Atom Bomb

The Allies were particularly worried that Germany might develop and use a fission weapon. The Germans had discovered the fission process and had a high level of scientific and technical skill.

Urged by Albert Einstein and others, President Franklin D. Roosevelt launched a research and development program known as the Manhattan Project, briefly outlined here. The ultimate objective of

The Manhattan Project: Research, Development and Production

Location	Activities
Columbia University, New York City	Gaseous diffusion uranium isotope separation
University of Chicago "Metallurgical Laboratory"	First nuclear reactor; materials and chemical studies
University of California, Berkeley	Electromagnetic uranium isotope separation; chemistry of plutonium
Oak Ridge, Tennessee	Production facilities; chemical process and reactor research
Hanford Works, Richland, Washington	Plutonium production reactor
Los Alamos, New Mexico	Weapons research
Iowa State College	Reactor materials

*The first account of the atom bomb project appeared in a 1945 book by H. D. Smyth (see References).

the whole enterprise was to harness the fission process to form an explosive, i.e., a bomb. The first material sought was highly enriched uranium, around 90 percent ^{235}U, starting with natural uranium, 0.7 percent ^{235}U (see the illustration). The second material was plutonium, the result of neutron bombardment of ^{238}U.

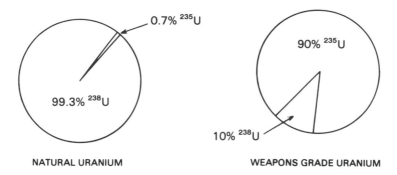

NATURAL URANIUM   WEAPONS GRADE URANIUM

Composition of different enrichments of uranium.

Two types of weapons in which the materials were to be used were devised at Los Alamos in New Mexico, under the direction of J. Robert Oppenheimer. In the "gun" type, sketched here, halves of the assembly were brought together in a supercritical* condition in a tube very quickly by the use of conventional chemical explosives. In the "implosion" type, also illustrated, a chemical explosive compressed the material to the supercritical state. In either case, a tremendous amount of energy was released almost instantly, with a great deal of radiation and heat. In the nuclear weapon, the chain reaction is used to cause rapid neutron multiplication through the consumption of fissile nuclei, as shown in the sketch.

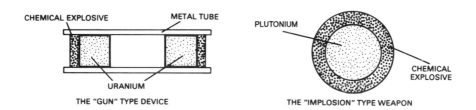

THE "GUN" TYPE DEVICE   THE "IMPLOSION" TYPE WEAPON

Nuclear weapons of the fission type.

---

*The terms "subcritical," "critical," and "supercritical" indicate whether a neutron in the chain reaction on the average produces neutrons in number less than one, one, or more than one.

The chain reaction. In just three steps one neutron is "multiplied" to form seven neutrons. Four uranium nuclei are fissioned in this example, each giving rise to energy.

The separation of ^{238}U from ^{235}U was needed to provide a readily fissionable material. Techniques tested were thermal diffusion, centrifuge, electromagnetic, and gaseous diffusion. Research on the electromagnetic isotope separation process was conducted at Berkeley under Ernest O. Lawrence. Large-scale facilities at Oak Ridge produced enough highly enriched uranium (about 90 percent ^{235}U) for one of the first bombs. Gaseous diffusion, however, proved to be the most economical process. It now provides the slightly enriched uranium (about 3 percent ^{235}U) used in present nuclear power reactors. Because of their construction and type of fuel, reactors cannot explode as does a bomb.

Separation of uranium isotopes by gaseous diffusion. The uranium-235 atoms pass through the "barrier" more readily than do the uranium-238 atoms. Thousands of such units are connected.

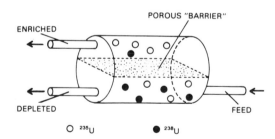

The World's First Nuclear Reactor

The first nuclear reactor was built at the University of Chicago in 1942 under the leadership of Enrico Fermi. The questions to be answered were, first, whether a controlled chain reaction involving neutrons and uranium could be achieved and, second, whether irradiation of uranium by neutrons could produce enough plutonium to build a bomb.

The reactor shown in the artist's sketch was constructed of chunks of natural uranium as metal and oxide embedded in graphite blocks. The graphite (carbon) served to slow the neutrons to low energy; that is, it served as "moderator." This first uncooled "pile" went critical on December 2, 1942, and reached a power of 200 watts.

Artist's sketch of Fermi's chain-reacting pile at the University of Chicago, 1942.

The success of the Chicago reactor led to the construction of several large plutonium production reactors at the Hanford Works, located near the present city of Richland, Washington. These reactors also used graphite and natural uranium, with the uranium metal canned in aluminum to protect it from the effects of water coolant. The pieces of uranium, "slugs," were pushed through the reactor after they were irradiated.

A chemical operation called "reprocessing" was then applied to recover the plutonium that had been produced by neutron irradiation of ^{238}U. The aluminum coating was removed by sodium hydroxide, and the uranium was dissolved by nitric acid. The chemical bismuth phosphate was then added to precipitate the plutonium, leaving uranium and fission products in solution. A separate process was required to remove the uranium.

One disadvantage of the process was that rather large volumes of solids were mixed with the fission products. At the time, of course, the object was to get plutonium, and the waste problem was given a much lower priority. As the result of reprocessing spent fuel, large volumes of the waste were accumulated during and since the war, as the nation stockpiled nuclear weapons for the Cold War with the U.S.S.R. These defense wastes exist and must be dealt with, regardless of the rate of production of wastes from commercial nuclear power.

Many excellent books have been written about the development and use of the first nuclear weapons. Appendix B lists a few of these.

CHAPTER 9

Defense and Development

After the end of World War II in 1945, the United States investigated ways to use nuclear energy for peaceful purposes and, at the same time, sought to maintain and extend the nation's weapons capability. Policies made and actions taken in the period since 1945 have shaped the present state of development of nuclear energy, with its accomplishments, opportunities, and problems.

National Laboratories for Research

To administer research and development on both peaceful and military applications of nuclear energy, Congress created in 1946 the U.S. Atomic Energy Commission (AEC). In that year the Atomic Energy Act of 1946 charged the AEC to provide materials for defense purposes and to carry out weapons tests, to further the successful application of fission for nuclear power, and to find ways to use the new radioactive materials for beneficial purposes.

Several "national laboratories" were designated to carry out related research and development. The first of these were Oak Ridge National Laboratory in Tennessee, Argonne National Laboratory near Chicago, Brookhaven National Laboratory on Long Island, and Los Alamos National Laboratory in New Mexico. Additional centers in later years include Lawrence Livermore National Laboratory and Lawrence Berkeley Laboratory near San Francisco, Sandia National Laboratories in New Mexico, Bettis Laboratory near Pittsburgh, Idaho National Engineering Laboratory near Idaho Falls, the Pacific Northwest Laboratory near Richland, Washington, and others.

Defense Projects

The national nuclear defense of the United States was supported by several weapons production facilities, administered by the Atomic Energy Commission. These included isotope separation plants using gaseous diffusion at Oak Ridge, Tennessee, at Paducah, Kentucky, and at Portsmouth, Ohio (pictured). These facilities have also supplied most of the fuel for research, test and commercial reactors. For the continued production of plutonium, the Hanford reactors (also pictured) were operated for several additional years.

The gaseous diffusion plant for separating uranium isotopes at Portsmouth, Ohio. (Courtesy of Battelle Memorial Institute.)

Plutonium-producing reactor facilities at Hanford. These were operated for the first time on December 17, 1944. (Courtesy of the Department of Energy.)

The plutonium generated in the AEC reactors is called "weapons grade" plutonium, containing a rather small concentration of the isotope plutonium-240. This isotope is undesirable because some of its nuclei undergo spontaneous fission, yielding neutrons, which tend to cause premature detonation and inefficient explosion in a weapon. In contrast, "reactor grade" plutonium, produced by long irradiation, has a large plutonium-240 content and is not useful for constructing a weapon.

For the generation of tritium, used as an ingredient of the thermonuclear weapon (hydrogen bomb or H-bomb), "production reactors" were built at the Savannah River Plant near Aiken, South Carolina. The exact nature of nuclear weapons and the stockpile of material is not publicly known, but the weapons capability at its peak included some 2000 launchers and 10,000 warheads. Part of the materials called defense wastes come from the chemical treatment (reprocessing) of spent fuel from the production reactors and from the fabrication of plutonium weapons.

A weapons testing program was carried out over the many years of the Cold War, to improve yield and to study military effects. Sites include Pacific bases at Bikini, Eniwetok, and the Marshall Islands, and especially at the Nevada Test Site near Las Vegas. Significant radioactive fallout was experienced, leading to a ban by treaty on aboveground tests.

The Nuclear Navy

The original production reactors were composed of graphite, with cylindrical natural uranium fuel rods. The availability of uranium enriched in ^{235}U made possible a new type of reactor, consisting of metal alloy fuel plates, with water as both cooling agent (coolant) and moderator. This reactor was investigated at Oak Ridge, Tennessee, as a possible power source for naval vessels, especially submarines. Between 1948 and 1953 a submarine reactor was built and tested at Idaho Falls, Idaho, by Argonne National Laboratory and Westinghouse Electric Corporation, under the leadership of Admiral H. G. Rickover.

The submarine *Nautilus* went to sea in 1955 powered by this nuclear reactor using enriched uranium fuel. Because the weight of fuel used was so small, the submarine was able to stay under water for months and to go under the ice at the North Pole. With its first core loading it went 62,000 miles.

Conventional submarines were replaced with nuclear-powered vessels because of the greater speed, freedom from noise, and ability to remain submerged for a long time. By cutting a reactor-powered submarine in two and inserting a new section, the first guided missile submarine was built. If necessary the rockets could deliver nuclear weapons thousands of miles away. The older Polaris missiles have been replaced with Poseidon, Tomahawk cruise missiles, and Trident ballistic missiles.

Few people realize the size of the active fleet of the U.S. nuclear Navy. It is constantly changing as new ships are built, others are overhauled or modernized, and still others are retired from service. At maximum strength there were around 150 submarines, 9 cruisers, and 7 aircraft carriers. The *Enterprise* shown in the photograph was the first of its type. It has eight nuclear reactors, 80 aircraft, and a crew of over 5000. The spent fuel from the naval reactors is sent to Idaho Falls for reprocessing.

The nuclear-powered aircraft carrier *Enterprise*. Einstein's formula relating mass and energy is spelled out by crew members. (Courtesy of the Office of Information, Department of the Navy.)

With the reduction in international tensions, decisions are being made about the number and character of nuclear naval vessels to be retained. Some new nuclear waste will be developed as ships are decommissioned.

Electrical Generation

The Atomic Energy Act of 1954 called for encouragement of nuclear reactor development for commercial electric power. Industrial organizations cooperated with the AEC in programs designed to learn what reactor types were suitable for economical production of electricity. The homogeneous reactor consisted of a water solution of a uranium salt, with both fuel and moderator circulated through heat removal equipment. Corrosion problems led to abandonment of this concept. A graphite-moderated reactor similar to the production types but cooled with liquid sodium was tested and found impractical. A reactor was tried that was cooled with a petroleum-derived liquid with boiling point much higher than that of water. The cooling agent turned out to be affected adversely by radiation.

Successful, however, was the Experimental Breeder Reactor, at Arco, Idaho (pictured). In 1951 it produced the first commercial nuclear electric power. It was of the fast-reactor type, without moderator, and cooled by liquid metal.

A breeder is a reactor that takes advantage of the number of neutrons, an average of about three, emitted in fission caused by fast neutrons in the element plutonium. Although one neutron must be used to continue the chain reaction, another is available to be absorbed by the fertile isotope uranium-238 to produce additional plutonium. More fuel can be produced than is burned to obtain power.

The advantage of the breeder is that it can use the abundant uranium-238 (converted into plutonium-239) as fuel while other reactors use mainly the scarcer uranium-235. The total uranium resources of the earth are available if the breeder is adopted. Japan is leading research on breeding, with lesser effort in the United Kingdom, France, and the United States. Studies of a complete system called the Integral Fast Reactor have been conducted at Argonne National Laboratory.

By the mid-1960s two commercial versions of what is called the light-water reactor had been developed. One is the pressurized water reactor, developed by Westinghouse Electric Corporation as a spinoff of the submarine program. The other is the boiling water reactor, first tested at Argonne National Laboratory and perfected by the General Electric Company. In the period 1965-1970, large numbers of reactors were ordered by U.S. utilities, and a major construction program was begun. By 1993 more than 100 power reactors were operating in the U.S. and almost three times as many abroad. The photo shows the exterior of a typical nuclear power plant, and the schematic drawing shows the basic system. One other major type of reactor, CANDU, with heavy water, containing deuterium as the moderator, is in operation in Canada, India, and Pakistan. Gas-cooled reactors have been used extensively in the United Kingdom, and the graphite-moderated Chernobyl type is found in the Soviet Union.

(Below Left) Side view of EBR-I showing the radiation shield.

(Above Right) The first use of electric power from atomic energy, December 20, 1951.

The Experimental Breeder Reactor (EBR-I) at Arco, Idaho. (Courtesy of Idaho Operations Office, Department of Energy.)

The H. B. Robinson nuclear plant at Hartsville, South Carolina, operated by the Carolina Power and Light Company. On the right is the reactor containment building; on the left is the turbine-generator. (Courtesy of George Zellars.)

A perspective of the present role of nuclear power in the U.S. can be gained from some statistics. First is the energy produced from different sources, shown in the following table. The energy goes to provide transportation (27.2%), residential and commercial (36.1%), and industrial applications (36.7%). As we see in the other table, the electricity produced by commercial nuclear power plants exceeded that by plants using natural gas, oil, and hydroelectric.

The average retail cost of electricity to all consumers in the U.S. is 6.9 cents per kilowatt-hour. From the production figure given in the table, we deduce the value of electricity from all sources to be around $200 billion per year.

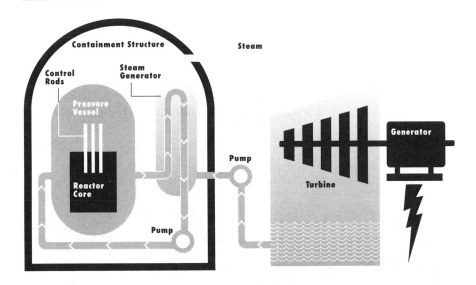

Flow of fluids in the pressurized water reactor (PWR). Water is heated by the nuclear fuel but kept under pressure so it will not boil. It is piped from the pressure vessel to a steam generator. There it transfers its heat to a second supply of water, which boils to make steam for the turbine. (Courtesy of Nuclear Energy Institute.)

Total Energy Produced in the U.S. by Primary Type
(From Department of Energy, see References)

Energy Source	Energy in Quadrillions of Btu*	Percent
Natural Gas and Liquids	21.38	32.5
Coal	20.49	31.1
Petroleum	14.48	22.0
Nuclear Power	6.52	9.9
Hydroelectric	2.71	4.2
Other	0.18	0.3
Total	65.81	100.0

*British thermal units. One Btu of heat energy will raise the temperature of a pound of water by one degree Fahrenheit.

Electrical Energy Produced in the U.S.
(From Department of Energy, see References)

Type of Plant	Energy in billion kWh	Percent
Coal	1639	56.8
Nuclear	610	21.1
Hydroelectric	269	9.3
Natural Gas	259	9.0
Petroleum	100	3.5
Other*	10	0.3
Total	2887	100.0

*Includes geothermal, wood, waste, wind, photovoltaic, and solar.

CHAPTER 10
Uses of Isotopes and Radiation

In accord with the Atomic Energy Act of 1946, the AEC encouraged the development of new applications of radioisotopes and radiation. The objective was to find beneficial uses that would provide direct economic return, advance scientific knowledge, and more generally improve human life. Uses were found in many areas, including medical diagnosis and treatment, agriculture, industry, research, and space travel.

A review of just a few applications of nuclear technology will provide historical background and set the stage for consideration of the management of low-level wastes, which are by-products of the use of isotopes as well as the operation of nuclear power plants.

Tracers

Complicated physical, chemical, and biological processes can be studied by the use of isotopes that "trace" the material of interest. For example, the radioisotope sodium-24, formed by neutron absorption in ordinary sodium-23, has a half-life of 15 hours. If a salt solution containing a small amount of sodium-24 is injected into a person's vein, the speed of blood flow through the body can be measured by the radioactive emanations.

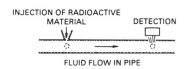

INJECTION OF RADIOACTIVE MATERIAL   DETECTION

FLUID FLOW IN PIPE

THE SPEED OF FLOW OF LIQUID IN A PIPE CAN BE FOUND BY TIMING THE ARRIVAL OF RADIOACTIVITY.

The flow of oil or other fluid in a long pipeline is measured by injecting a small amount of radioisotope at one end and detecting its passage at the other end, as shown in the sketch.

The isotope phosphorus-32, half-life 14.3 days, can be mixed with the fertilizer applied to the roots of plants, as sketched here. The radioactive substance is then taken up by the plant,

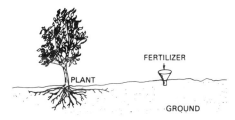

FERTILIZER

PLANT

GROUND

UPTAKE OF FERTILIZER CAN BE GAUGED BY MEASURING THE ACTIVITY OF PHOSPHORUS-32 IN PLANT LEAVES.

Two uses of radioactive tracers.

and its presence in the stem and leaves reveals where the fertilizer goes. Effects of placement, timing, and amounts of fertilizer can be deduced using this method.

If a radioactive species of an element is used in a chemical compound, the substance is said to be "labeled," i.e., given a special identification. Tritium (hydrogen-3), as an isotope of hydrogen with a half-life of 12.3 years, and radiocarbon (carbon-14), half-life 5715 years, are excellent isotopes for labeling organic compounds (which contain carbon, hydrogen, oxygen, and nitrogen) in a great variety of biological research studies. Many molecules have carbon atoms in different places, so that several differently labeled compounds can be formed.

Imaging

Nuclear medicine is a recognized specialization that uses nuclear techniques to diagnose ailments. Radioactive materials are used in a process called imaging. The patient is given a solution containing a small concentration of an isotope that has a special affinity for tissue where a difficulty is suspected. A gamma-ray-sensitive radiation detector then scans that part of the body. The pattern of measured radiation reveals information about size, shape, and condition of the organ. The method is widely employed for certain infections, gall bladder and heart diseases, and bone cancer. The table and the sketch indicate how this technique works.

Radioisotopes Used in Imaging of Organs of the Body

Iodine-125, 131	Thyroid, liver, kidney, heart, lung, brain
Chromium-51	Spleen
Technetium-99m	Brain, liver, spleen, kidney, lung, bone
Selenium-75	Pancreas
Strontium-85	Bone

Locating a brain tumor by imaging—injecting a radioactive chemical. The arrow near the top of the head points to the malignancy. (Courtesy of Robert T. Morrison, M.D., Vancouver, British Columbia, General Hospital.)

A special type of nuclear imaging for research goes by the acronym PET (positron emission tomography). One injects a compound containing a positron emitter such as fluorine-18. Combination of the positron with an electron yields two gamma rays.

Therapy

The treatment of disease by the use of radiation from radioisotopes is a common practice. Some of the diagnostic isotopes can be used with higher activity levels to irradiate organs. For example, radiation from phosphorus is applied in leukemia therapy, and that from iodine can treat hyperthyroidism or brain tumors. Several iodine isotopes are available: iodine-125, half-life 59 days, comes from the cyclotron, a particle accelerator; iodine-131, half-life 8 days, is a fission product.

The radiation from cobalt-60, half-life 5.3 years, consists of two gamma rays with an average energy 1.25 MeV. These rays continue to be used widely in the world for the treatment of cancer, even though there is an increasing application of accelerator-produced radiation.

Implantation of a capsule of radioactive material in an organ provides local irradiation. In earlier times, radium was used; currently popular isotopes are iridium-192 (74 days) and iodine-125 (59 days).

An especially valuable radioactive pharmaceutical for therapy is strontium-89 (50.5 days), which goes to the bone and alleviates intense pain associated with metastatic bone cancer. Other shorter-lived substances being tested in a program of the International Atomic Energy Agency are rhenium-186 (3.8 days) and samarium-153 (1.9 days). Generally, beta or alpha emitters are preferred to gamma emitters because of the greater localization of energy deposition.

Pharmaceutical Research

The development of new drugs for curative purposes by pharmaceutical companies requires extensive use of radioactive materials. Regulations of the U.S. Food and Drug Administration (FDA) call for test data on the behavior of a proposed drug. Tests with animals and humans are required. Tracer studies employ tritium and carbon-14, but also involve a variety of other isotopes, including phosphorus-32 (14 days), chromium-51 (28 days), iodine-125 (59 days), sulfur-35 (87 days), cobalt-57 (272 days), and chlorine-36 (3×10^5 yr), the only long-lived isotope of chlorine. Several of these have half-lives long enough that residues form a waste requiring disposal rather than holding for decay.

When a drug itself is radioactive, data on the absorbed dose to specific organs and the whole body must be supplied to get FDA approval.

Research reactors serve as the principal source of radioactive materials around the world. In the U.S., neutron-induced radionuclides are supplied by Canada; charged particle-induced isotopes are produced by local cyclotrons.

Radiography

X-rays have been used in medical diagnosis for many decades. A refinement in radiography is CT (computed tomography), in which x-ray scattering data are processed by a computer to give very precise information about thin layers of the body. Modern diagnosis involves a combination of nuclear and non-nuclear methods, including nuclear medicine, PET, CT, ultrasonics, and magnetic resonance.

Radiography is also applied in industry for the inspection of metal parts for flaws, as shown in the sketch. Cobalt-60 gamma rays are preferable to x-rays. Advantages are (1) the high energy for good penetration of metal and (2) portability, achieved without the need for an electrical supply, as an x-ray machine requires. The method is especially good for checking welded metal joints in a nuclear reactor cooling system and for locating areas of metal fatigue in aircraft structures.

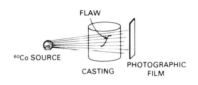

FLAW IN A LARGE METAL CASTING BEING DETECTED BY COBALT GAMMA RAYS.

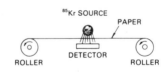

THICKNESS OF PAPER BEING MEASURED DURING MANUFACTURE.

Gauging

The rays from radioisotopes are useful for making a variety of measurements. One example is the continuous testing of the thickness of paper during its manufacture (as sketched

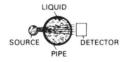

DENSITY OF A LIQUID FLOWING IN A PIPE BEING FOUND BY OBSERVING GAMMA RAY INTENSITY.

Industrial uses of radiation.

on this page) by the use of beta particles from krypton-85. Variations are detected in the number of particles that get through the paper. In soils the moisture content is measured by the migration of neutrons coming from the reaction of alpha particles on beryllium. Also, as pictured, the density of a fluid moving in a pipe can be found without taking a sample by detecting the gamma rays that get through the pipe. Similarly, the level of liquid in cans of soft drink on a conveyor belt can be checked, at a rate of 2000 cans per minute.

Dating

The age of archeological and historical objects is found by the carbon-dating technique. There has always been a certain amount of carbon-14 in the atmosphere. Plants such as trees use CO_2, and traces of this 5715-year half-life isotope are deposited in their tissues. At any later time, measurement of the carbon-14 content tells us the age of any artifact made from the plant.

The age of mineral deposits can be found by examining the ratio of uranium and lead isotopes, ^{238}U and ^{206}Pb. (Recall the uranium radioactive decay chain given on p. 14.) A recently deposited mineral would have little lead; an ancient one would have a great deal. The age of the earth has been estimated by this method.

Neutron Activation Analysis

The absorption of neutrons by many stable isotopes renders them radioactive, and detection of the resulting radiation indicates the amount and type of the original substance. Extremely minute amounts can be measured by this process called "neutron activation analysis." Examples are: industrial mercury pollution in water that is taken up by animals and deposited in their tissues; crime investigation, in which the composition of stolen goods is compared with the original stock; authentication of old paintings by testing the paint for agreement with that available in earlier times; measuring an alloy for minute traces of undesirable elements. An explanation of the extinction of dinosaurs comes from measurement of the iridium content of layers in seabed sediments. It is proposed that a collision of an asteroid with the earth created dust that reduced sunlight needed by plants on which the dinosaurs depended.

Elimination of Pathogens

Radiation is used to kill pathogens (disease-causing agents such as bacteria, fungi, and viruses). Foods can be kept from spoiling by the use of gamma rays from cobalt-60 or cesium-137, without the use of chemicals that may be carcinogenic (cancer-causing). The limited doses applied kill organisms but produce no radioactivity or harmful chemicals. Treatment can suppress trichina in pork and salmonella in chicken and reduce the number of food-borne illnesses and deaths. Food irradiation is practiced in a number of countries around the world, with strong recommendation by the World Health Organization for its use as a safe way to reduce food losses due to insects and spoilage. Only recently has food irradiation been applied in the U.S., even though the Food and Drug Administration has given approval for irradiation of many food products. Spices and strawberries are the principal commodities commercially irradiated.

Medical supplies and pharmaceuticals are sterilized by gamma irradiation. Contamination is prevented by irradiating equipment and materials within sealed plastic containers.

Blood used for a transfusion to a patient undergoing an organ transplant is treated with radiation from radioactive substances to reduce the number of white blood cells and thus prevent attack and rejection of the organ.

Elimination of Insect Pests

Certain insect pests can be controlled by using a special radiation technique. An example is the screwworm fly, whose larvae can kill cattle; it was recently eradicated in the southern part of the U.S. Large numbers of males of that species were sterilized by gamma irradiation and then released. Since their mating resulted in no off-spring, the screwworm population dropped rapidly. Emergency action in Libya was taken in 1988-1990 upon discovery of a screwworm infestation. A "fly factory" in Mexico supplied millions of sterilized males, which were dropped by plane like crop dusting. Within five months, the pest was eliminated, thus protecting African wildlife.

Special Power Sources

The heat produced by the beta-decay of strontium-90 is useful for thermoelectric generators, which produce small but steady, reliable amounts of electric power for remote locations. One navigational beacon powered by strontium-90 was in continuous operation for 16 years.

The isotope plutonium-238, half-life 87.7 years, is produced by neutron irradiation in a nuclear reactor. The isotope is an alpha emitter (5.5 MeV) that is used in thermoelectric power generators for space missions. Equipment for the exploration of the moon by Apollo and of other planets by Pioneer, Viking, and Voyager has been powered by such radioisotope devices. Voyager 2 started with 470 watts of electric power delivered by a radioisotope thermoelectric generator (RTG). It visited Jupiter, Saturn, Uranus, and Neptune, sending photographs back to earth by radio signals. The final goal for exploration of the solar system is a flyby of Pluto, over three billion miles from earth. This most distant planet has an unusual eccentric orbit and rotates with its large moon, Charon, about their center of gravity. An RTG is used instead of solar panels because sunlight is so weak at Pluto's distance from the sun, 40 times that of Earth. It thus receives less than one percent as much light as the Earth.

Voyager **spacecraft with radioisotope thermoelectric generator as power source.**
(Courtesy of the National Aeronautics and Space Administration.)

The spacecraft *Galileo*, equipped with a higher power RTG, provided pictures in 1994 of the collision of the Shoemaker-Levy comet fragments with the planet Jupiter.

Smoke detectors for fire protection use a heavy artificial radioactive isotope, americium-241, half-life 432 years. The principle of operation is described by one of its manufacturers, Fenwal, Inc. The detector consists of an ionization chamber with a small radioactive source of 0.7 μCi, sealed in stainless steel. The source continuously emits alpha particles to ionize the air and allow a small current to flow between two plates. Combustion products from a fire enter the chamber and change the current and voltage. An associated electric circuit produces the alarm. This type of smoke detector responds very quickly to a broad range of particle sizes.

Resultant Wastes

Each application of isotopes involves a certain amount of radioactive waste. Some arises in producing the isotopes and in fabricating the devices; some remains after completing an experiment, or a medical test, or at the end of the useful life of a piece of equipment. The degree of hazard resulting from the radioactivity depends on many factors—the amount of material, the rate at which it decays, the energy and type of radiation, whether the substance can enter the human body, and how long it remains in the body. Obviously, an isotope of half-life 15 hours, such as sodium-24, will be essentially gone after a week's storage, at which time the material can be discarded.

Others such as 5.3-year cobalt-60 require precautions over a longer time span.*

The foregoing account reveals the ironic fact that the very radiation that can harm human beings can also benefit health and general welfare.

People often ask why we do not use all the radioactive wastes for beneficial purposes and thus reduce the need for storage and disposal. The answer is related to supply and demand, and thus economics. There is an abundance of radioactive materials in spent reactor fuel, but chemical recovery and purification is required to extract the isotopes. The demand for radionuclides as yet does not warrant large-scale processing. The wastes could be a source of heat, but the cost of preparing devices and of providing radiation protection would be prohibitive except for special applications. In summary, there are more radioactive substances than we currently can use economically.

---

*A good discussion of both the uses of radiation and its effects is found in Eric J. Hall, *Radiation and Life*, 2nd ed., Pergamon Press, 1984.

CHAPTER 11

Classification of Wastes

Nuclear materials comprise a great variety of isotopes, elements, chemical compounds, and mixtures. Definitions are important because substances are managed by law and regulation according to classification.

Radioactive materials fall into several categories according to their origin, the type of material present, and their level of radioactivity. The first and broadest distinction is:

Defense

Commercial.

Defense Wastes

Defense wastes have been generated over the period during and since World War II, at three main Department of Energy (DOE) installations—the Hanford Site near Richland, Washington, Idaho National Engineering Laboratory, near Idaho Falls, Idaho, and the Savannah River Plant near Aiken, South Carolina. Plutonium and other isotopes were separated from production reactor spent fuel at Hanford and at Savannah River, while naval propulsion reactor spent fuel was processed at Idaho Falls. In each case the chemical process has left a residue of fission product waste. Other plutonium-contaminated wastes have evolved from weapons fabrication at Rocky Flats, Colorado, and several other sites.

Commercial Wastes

Commercial wastes are those produced by reactors used for electrical power, by facilities used to process reactor fuels, and by a variety of institutions and industries. There is only a small volume of commercial wastes from reprocessing because most of the fuel from power reactors has been left in the form of irradiated fuel assemblies, the spent fuel. Used fuel remains highly radioactive for years after it is removed from the reactor.

The only reprocessing of U.S. commercial wastes was by Nuclear Fuel Services, Incorporated, at West Valley, New York, in the period

1966 to 1972. This plant was shut down because it was uneconomical to operate.

Since 1972 spent fuel has been accumulating at nuclear power plants. As more reactors came on line, the rate of growth increased. The total activity at present greatly exceeds that of the earlier by-products of reprocessing.

Comparisons and contrasts among the principal wastes can be noted. Both defense and commercial reactor wastes stem from the operation of fission reactors, but defense wastes generally are considerably older and less radioactive. Chemical separations used during World War II were designed to extract plutonium well but not to minimize residual wastes. Even though there is a small volume of fission products in defense wastes, the chemicals with which they are mixed add considerable volume. For all practical purposes, there are no separated commercial reactor wastes, but the volume of spent fuel is large because of uranium and metal structures. Institutional and industrial wastes generally have a low radioactivity level since they do not usually contain fission products. The volume is rather large, however.

Three Important Types of Wastes

Another distinction among radioactive wastes is:

High-level

Transuranic

Low-level.

High-level wastes (HLW) are those resulting from reprocessing of spent fuel or are the spent fuel itself, either of defense or commercial origin. When spent fuel is chemically processed, the residue consists of fission products and small amounts of plutonium isotopes. Such HLW and spent fuel are the candidates for disposal by burial deep in the ground. We shall discuss such geologic disposal in a later section.

Transuranic wastes (TRU) are those containing isotopes above uranium in the periodic table of chemical elements. They are the by-products of fuel assembly and weapons fabrication and of reprocessing operations. Their radioactivity level generally is low, but since they contain several long-lived isotopes, they must be managed separately. This classification is composed of isotopes with half-lives greater than 20 years and giving a total activity of greater than 100 nanocuries* per gram of waste material. Isotopes include plutonium-239, half-life

---

*Recall that one nanocurie is 37 disintegrations per second.

2.411 x 10^4 year; americium-241, 432.2 year; americium-243, 7.37 x 10^3 year; curium-244, 18.11 year; and curium-245, 8.5 x 10^3 year. A very long half-lived isotope is also present—neptunium-237, 2.14 x 10^6 year.

Transuranic wastes give off very little heat, and most of them can be handled by ordinary methods not requiring remote control. For many years they were buried in shallow trenches, but since 1970 they have been placed in retrievable storage. Some plutonium-contaminated soil at Rocky Flats in Colorado, resulting from fires and leaks, is being cleaned up.

Mill tailings are the residue from the physical and chemical processing of uranium ore to obtain uranium. Tailings are only slightly radioactive, but there is an enormous volume of them. We will discuss this special material in a later chapter.

Low-level wastes (LLW) are officially defined as all wastes other than those defined above. The bulk of LLW have relatively little radioactivity and contain practically no transuranic elements. Most of them require little or no shielding, may be handled by direct contact, and may be buried in near-surface facilities. Part of the LLW, however, have high enough radioactivity that they must be given special treatment and disposal. Low-level wastes come from certain reactor operations and from many institutions such as hospitals and research organizations and from industry. LLW are divided by the NRC into three classes—A, B, and C, generally in increasing degree of radioactivity and corresponding need for secure disposal. Those LLW that have even higher activity are called "greater-than-Class-C" wastes, considered to have a hazard comparable to that of spent fuel.

Mixed wastes are low-level wastes that contain both hazardous chemicals and radioactive substances. Hazardous wastes are defined as materials that are toxic, corrosive, inflammable, or explosive. They contain specific elements such as lead and mercury, pesticides such as DDT, and cancer-producing compounds such as PCBs. In Chapter 19 we will discuss the dual regulation of these wastes by EPA and NRC.

There is no standard accepted scheme for listing types of wastes. Other waste classifications are found. For example, in the Department of Energy's environmental impact statement on commercial wastes, one finds these categories, each applied to the word "wastes": primary, secondary, nuclear power plant, spent fuel basin storage, fuel reprocessing plant, mixed-oxide fuel fabrication, decommissioning. In an official tabulation of waste inventories prepared by Oak Ridge National Laboratory for DOE, one finds these major classes: commercial spent fuel, high-level waste, transuranic waste, low-level waste, uranium mill tailings from commercial operations, environmental restoration wastes, commercial decommissioning wastes, and mixed low-level waste. Part of the wastes now classified as "environmental restoration" were previously identified as "remedial action," related to cleanup actions at DOE sites.

Amounts of Defense Wastes

A large volume of wastes classified as high-level defense waste is now stored in underground tanks and bins at three main government sites—Hanford, Idaho Falls, and Savannah River. Estimated amounts according to physical form are shown in the next table. In addition, there is a large accumulation of transuranic wastes at DOE facilities, also listed in a table here. About an eighth of the TRU could be retrieved, at considerable expense. Not included in the table is a large volume of soil that is slightly contaminated with plutonium.

**Existing Department of Energy High-Level Wastes,
in Thousands of Cubic Meters***

Location	Liquid	Sludge	Salt Cake	Slurry	Misc.	Capsules	Total
Savannah River	59.3	14.3	53.1	—	0.2	—	126.9
Idaho Falls	7.7	—	—	—	3.5	—	11.2
Hanford	25.1	46.0	93.0	94.7	—	0.00353	258.7
Total	92.1	60.3	146.1	94.7	3.7	0.00353	396.8

*From *Integrated Data Base for 1993: U.S. Spent Fuel and Radioactive Waste Inventories, Projections, and Characteristics,* DOE/RW-0006, Rev. 9, March 1994. "Liquid" is the acidic by-product of reprocessing; "Sludge" comes from neutralization of acidic solution; "Salt Cake" is from evaporation; "Slurry" is waste in double-shell tanks; "Misc." includes calcines and precipitates; "Capsules" are special containers of separated strontium-90 and cesium-137.

**Transuranic Wastes at DOE Facilities,
in Thousands of Cubic Meters***

Location	Buried	Stored
Hanford	109.0	15.7
Idaho Falls	125.7	64.8
Los Alamos	—	10.6
Oak Ridge	10.6	2.2
Savannah River	4.5	10.0
Rocky Flats	—	1.0
Nevada Test Site	—	0.6
West Valley	—	0.5
Other	1.0	0.4
Total	249.8	105.9

*DOE/RW-0006, Rev. 9, March 1994.

It is difficult for us to visualize very large quantities such as a million barrels of oil, a billion dollars, and a million cubic feet. To appreciate the total volume figure of 394,800 cubic meters cited in the table, merely note that this is the same as a cube about 241 feet on a side, i.e., sitting on and covering a little over a square acre of land, which is 208.71 feet on a side.

The Hanford plutonium project is seen to have the largest volume of defense wastes. The site was selected immediately after Fermi's successful test of the first nuclear reactor. The semiarid area in southeastern Washington State (see the two maps) was selected because of its remoteness and the availability of water from the Columbia River. During World War II and afterward, the federal government constructed there a total of nine plutonium production reactors, five chemical processing plants, and 177 storage tanks. Of these tanks, 149 were single-walled, constructed between 1943 and 1964; 28 were double-walled, built between 1968 and 1986. Many of the single-walled tanks have leaked as a result of deterioration of the metal with age, plus inadequate attention. Some million gallons are estimated to have gone into the ground. Fortunately, the site for the tanks was chosen in part for the impervious nature of the soil, and the radioactive material has not migrated very far.

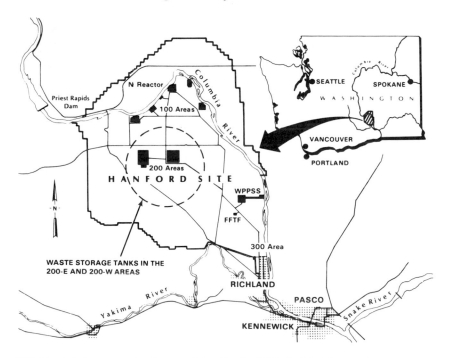

High-level defense waste storage at the Hanford Site in Washington State.

In 1990, one of the tanks started to generate hydrogen gas that would vent every 100 days. Concern arose about the possibility of an explosion that would spread radioactivity. To stir the waste and prevent buildup of large amounts of hydrogen, a mixer pump was developed, and installed in the tank, where it has undergone testing. In another set of tanks, the chemical ferrocyanide had been introduced to cause cesium to settle to the bottom. There is a potential for explosion in these tanks if the temperature gets too high, and monitoring devices have been installed.

Throughout the period of World War II and the Cold War, government emphasis had been placed on production of weapons material, and the growing waste problem was neglected. As a consequence, information about the contents of the tanks is inadequate. The investigation of composition and condition must precede remedial action, but it will also generate new waste streams to be managed. Meanwhile, the facilities must be maintained in a safe condition.

In a later chapter, we will address the major challenge of the long-term environmental restoration program for Hanford and the other DOE sites. This project parallels the other two national tasks—(a) building a suitable repository for high-level wastes and (b) developing sites for the disposal of low-level radioactive wastes.

Other Nuclear Materials

Certain other materials requiring physical security, transportation, or disposal are classified and defined as follows:

(a) "Source materials" are uranium or thorium or their ores containing as much as 1/20 percent of those elements. They serve as the source of fissionable material.

(b) "Special nuclear materials" (SNM) are uranium enriched in the isotope uranium-235 or manmade fissile elements such as plutonium or uranium-233. They are special because they can be reactor fuel or weapons.

(c) "By-product material" is radioactive material (excluding SNM) produced by irradiation in reactors or is the residue from the extraction of uranium from ore. It is a by-product from another objective, nuclear power.

CHAPTER 12

Spent Fuel from Nuclear Reactors

The main reactor type in use in the U.S. and throughout the world is the light-water reactor (LWR). It is so named because it uses ordinary water formed from hydrogen (not deuterium, as in a heavy-water reactor). The water serves as a "moderator," the substance composed of light elements with which neutrons collide and slow down. The water also serves as a "coolant," the medium that removes the fission heat. Two types of LWRs are in use: the pressurized-water reactor (PWR), in which the water is at high pressure and temperature but does not boil, and the boiling-water reactor (BWR), in which steam is produced directly in the reactor by limited boiling at relatively low pressure. Refer again to the coolant flow diagram for the PWR on p. 51.

Reactors in the U.S. and Abroad

The 116 commercial reactors in the U.S. are all of the light-water type. Two-thirds are PWRs, one-third are BWRs. As of the middle of 1994, 109 power reactors were in operation and seven others under construction, scheduled for completion. Some sites have more than one reactor: for example Duke Power Company operates three reactors—Oconee-1, -2, and -3 near Seneca, South Carolina.

The total power capacity of the 116 U.S. reactors is around 108,000 megawatts (108 GW). If all were in operation at achievable capacity factors, they would provide about 25 percent of the nation's electricity production.

The use of nuclear power abroad continues to grow. The following table shows the status of nuclear plants worldwide. A total of 30 countries have operable reactors and three more have reactors under construction. As seen in the table, the five nations that have the greatest nuclear power capacity in operation are the U.S., France, Japan, Russia, and Germany.

Nuclear Fuel Irradiation

Modern reactors use uranium that has a higher percentage of ^{235}U (3%) than is found in nature (0.7%). The fuel is in the form of uranium dioxide UO_2 as small pellets about 3/16 inch in diameter and

World Nuclear Power Plants, Operating and Total, as of June 30, 1994
(Adapted from *Nuclear News*, American Nuclear Society, September 1994)

Country	Number of Reactors		Megawatts Electrical	
	Operating	Total	Operating	Total
Argentina	2	3	935	1627
Belgium	7	7	5527	5527
Brazil	1	3	626	3084
Bulgaria	6	6	3420	3420
Canada	22	22	15,442	15,442
China	2	5	1800	3300
Cuba	0	2	0	834
Czech Republic	4	6	1632	3412
Finland	4	4	2310	2310
France	56	61	57,623	64,033
Germany	21	21	22,703	22,703
Hungary	4	4	1729	1729
India	9	16	1834	3874
Japan	47	54	36,946	43,692
Kazakhstan	1	1	135	135
Korea	9	14	7220	13,083
Lithuania	2	2	2760	2760
Mexico	1	2	654	1308
Netherlands	2	2	507	507
Pakistan	1	2	125	425
Philippines	0	1	0	605
Romania	0	5	0	3100
Russia	25	29	19,799	23,174
Slovakia	4	8	1632	3296
Slovenia	1	1	620	620
South Africa	2	2	1840	1840
Spain	9	15	7085	12,832
Sweden	12	12	10,002	10,002
Switzerland	5	5	2985	2985
Taiwan	6	6	4884	4884
Ukraine	14	20	12,095	17,795
United Kingdom	34	35	11,540	12,728
United States	109	116	99,510	107,994
Non-U.S.	313	378	236,410	287,066
Total	422	494	335,920	395,060

3/8 inch long, as pictured here. These are inserted into 14-foot-long, thin-walled (0.025 in.) tubes composed of an alloy of the element zirconium. Since the metal tube surrounds the fuel, it is often called "cladding"—a coating that prevents radioactive fission products from getting into the cooling water. The tube also provides support for the fuel. After the pellets have been introduced, the ends of the tubes are sealed to prevent water from getting in and gaseous fission products from getting out. Bundles of about 200 of the resulting fuel rods are

Fuel pellets for a pressurized water reactor. The uranium oxide contains ^{235}U at 3 percent enrichment. (Courtesy of AgipNucleare.)

formed, and the necessary space is maintained between rods as shown in the diagram. These fuel assemblies are about 8 inches on a side, are 14 feet long and weigh about 1200 pounds, but are readily handled with suitable hoists and cranes.

About 180 of the assemblies are closely packed vertically into what is called the reactor core, located in the lower part of the reactor vessel, pictured in the diagram. This vessel has a thick steel wall to

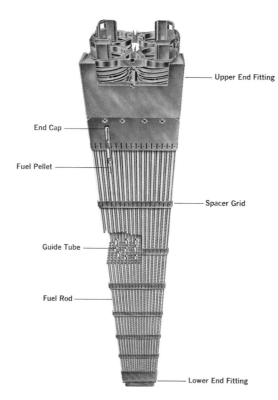

Upper End Fitting

End Cap

Fuel Pellet

Spacer Grid

Guide Tube

Fuel Rod

Lower End Fitting

Nuclear reactor fuel assembly. Bundles of 200 fuel rods, 14 feet in length, 8 inches on a side, are formed into an assembly. (Courtesy of Babcock and Wilcox Company.)

withstand the high pressure, 2200 pounds per square inch, resulting from operation with water at 600°F. Cooling water enters the vessel through large pipes welded to the side, comes down around the outside of the core, and is forced up past the fuel tubes, removing the heat of fission and keeping the rods at a reasonable temperature. As fuel is consumed, the reactor is held at a steady power level by adjusting the concentration of the neutron-absorbing boric acid in the cooling water. To change the power of the reactor or to shut the reactor down, the position of special metal control rods is adjusted.

Each typical operating period of a reactor is 1 year. At the end of that period, the operators

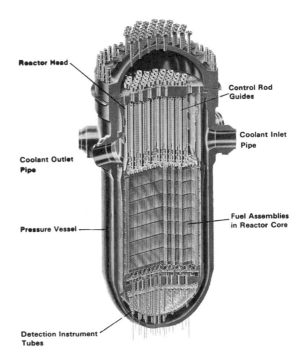

Reactor vessel of a pressurized water reactor. Cooling water enters the side, goes down the inside wall of the vessel, then comes up through the core containing uranium. (Courtesy of Babcock and Wilcox Company.)

take the top of the reactor vessel off and remove about one-third (60) of the assemblies on the basis of neutron exposure. This highly neutron-irradiated fuel is called "spent fuel" since it cannot sustain a chain reaction. However, it still contains some of the original ^{235}U and most of the ^{238}U. The remaining two-thirds are rearranged to give the best power, and a fresh one-third is inserted. The vessel is then closed and operation begins again.

New Isotopes in Used Fuel

During the 3 years that the fuel has been in the reactor, the irradiation of assemblies by neutrons has consumed some of the uranium and produced some new material. The content of uranium-235 as the main fissile material has been reduced from about 3 percent to about 1 percent, while the uranium-238 has gone down from 97 percent to 94 percent. A new fissile isotope has been produced by a neutron irradiation. Capture of a neutron in uranium-238 followed by beta decay leads to plutonium-239. This isotope complements the uranium-235 as fuel. It has a half-life of around 24 thousand years, emitting an alpha particle of 5.1 MeV energy. Other isotopes of elements above uranium in the periodic table (transuranic elements) are also produced. By successive neutron absorption ^{239}Pu becomes ^{240}Pu, fissile ^{241}Pu, ^{242}Pu, and short-lived ^{243}Pu. The following chart shows the "before" and "after" composition of spent fuel.

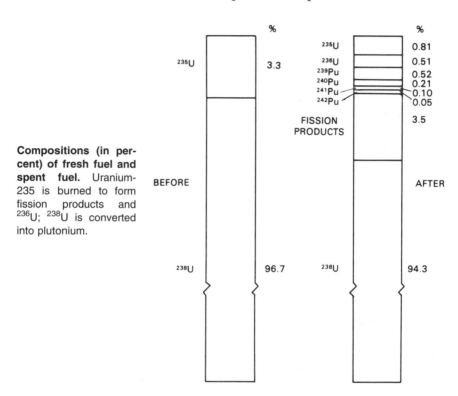

Compositions (in percent) of fresh fuel and spent fuel. Uranium-235 is burned to form fission products and ^{236}U; ^{238}U is converted into plutonium.

The spent fuel still has most of its original ^{238}U and a fairly high fissile fuel content. The total of ^{235}U, ^{239}Pu, and ^{241}Pu is 1.43 percent. The fuel removed is not really as "spent" as it might seem. The light-water reactors are called "converters" because they transform some of the ^{238}U into plutonium isotopes.

CHAPTER 13

Storage of Spent Fuel

Radiation and Heat from Spent Fuel

The spent fuel taken from the reactor after it has operated for a year is highly radioactive. The potential radiation dosage at contact with the surface of the fuel assemblies then is millions of rems per hour. Recalling that the lethal dose is around 400 rems, we can see that the fuel must be handled with great care. Moreover, it would be difficult to steal spent fuel without experiencing serious personal risk.

The spent fuel continues to be a source of heat and radiation after removal from the reactor, and thus is stored under water in a deep pool at the reactor site. The water keeps the fuel assemblies cool, and it acts as a shielding material to protect workers from gamma radiation. The water is kept free of minerals that would corrode the fuel tubes.

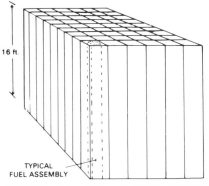

16 ft.

TYPICAL
FUEL ASSEMBLY

PERSPECTIVE VIEW OF METAL RACK, LOCATED IN A WATER POOL.

Fuel Storage Methods

The fuel is no longer suitable for operation in a reactor, but precautions must still be taken to avoid accidental criticality. The assemblies are kept separated in the pool by metal racks that leave about 1 foot between centers. This grid structure is composed of metal containing boron, which absorbs neutrons and prevents their multiplication. The diagrams depict these fuel storage racks, and the photograph shows a spent fuel pool.

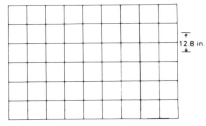

12.8 in.

TOP VIEW OF THE 6 by 9 ARRAY OF OPENINGS FOR SPENT FUEL.

Spent fuel storage rack to hold 54 fuel assemblies.

Water pool for storage of spent fuel at the Morris, Illinois, facility. (Courtesy of General Electric Company.)

Electric utilities that built reactors in the 1960s and early 1970s expected to send their spent fuel to a reprocessing plant within a few months after its removal from the reactor. Indeed, the Barnwell Nuclear Fuels Plant was constructed at Barnwell, South Carolina, for this purpose. There uranium and plutonium were to be extracted and recycled, with the fission product wastes stored for "cooling" and ultimate disposal.

However, a 1977 federal moratorium on reprocessing was instituted, requiring the utilities to keep the spent fuel at the reactor site. The requirement to store more fuel than anticipated was met by building closer-packed racks, reducing the spacing between fuel assemblies from 20 inches to 12 inches. Storage in existing pools is only a temporary measure because the spent fuel keeps coming—about 60 assemblies a year for each operating reactor. As discussed later, the federal government through DOE will take spent fuel for eventual disposal starting after the turn of the century. Meanwhile, however, U.S. utilities are responsible for the fuel. Some companies have transferred fuel from an operating plant to the pool of one under construction. Emergency storage space at national laboratories is quite limited, requiring that other solutions to the storage problem be found. One choice would be to build additional pools at individual reactor sites, at considerable expense because of requirements for a special cooling system, for a water purifer, and for stability against earthquakes. An away-from reactor concept would involve a large storage pool that serves several nuclear plants.

Alternative Storage Methods

To pack more fuel into existing pool space is preferable to constructing new pools. One way to increase capacity is to consolidate fuel. As tested by Duke Power and Westinghouse, fuel assemblies are remotely taken apart and tighter bundles are formed. Twice as much fuel can be put in the same place, and the process can be carried out without damage to the fuel. Consolidation is a satisfactory technique if the pool structure can support the extra weight.

Dry storage of spent fuel has the advantage of avoiding the need for new water pools. Containers can be procured as needed, and little maintenance is required. Virginia Power Company's dry cask demonstration facility at their Surry plant may be a model for many other utilities awaiting permanent disposal capability. The program of design, development, and testing involved many other organizations: A German firm, General Nuclear Systems, designed and built the casks; the Electric Power Research Institute (EPRI) provided financial support; DOE, EG&G Idaho, Bechtel, and Pacific Northwest Laboratory (PNL) did the testing at Idaho National Engineering Laboratory (INEL); and NRC licensed the facility. Design and safety evaluations included radiation levels and the effects of temperature, wind, tornado, fire, lightning, snow and ice, earthquake, and aircraft crash.

The body of the CASTOR V/21 cylindrical cask is cast iron 16 feet high and 7.9 feet in diameter, with 14.9 inches wall thickness. Fins on the outside help remove decay heat and maintain a safe cladding temperature. The cask will hold 21 PWR fuel assemblies in a stainless steel basket to provide proper separation. The total loaded weight is 120 tons.

The empty cask is lowered by special crane into the fuel pool, and assemblies are moved individually from the storage area and inserted in the cask. Double-sealed stainless steel lids are bolted on the top. Water is drained out and vacuum is applied to dry the fuel completely; then chemically-inert helium is introduced under pressure to protect the cladding against corrosion. The cask is lifted out and put on an A-frame on wheels, which is pulled by truck to the storage area, where the cask is unloaded on a concrete pad 3 feet thick, as shown in the photo. Eventually 84 casks on three pads will be installed. The cost of $1.2 million per cask is much lower than the cost of comparable pool storage.

Array of spent fuel dry storage casks of the Surry Power Station. (Courtesy of Virginia Power Company.)

Another design, by Westinghouse, holds 24 assemblies and discharges 12.6 kW of heat with the fuel rod surface temperature no higher than 140°F. The radiation dosage on the outside of the storage containers is 70 mrems/hr. Carolina Power and Light Company has built some large concrete storage modules that accept horizontal fuel casks. A hydraulic system is used to move the casks.

When DOE starts accepting spent fuel, it will require some storage space. One concept is the Monitored Retrievable Storage (MRS), a large facility located geographically between the generating companies and the fuel disposal site. The fuel would be repackaged at the MRS for disposal. Some devices that could be used for storing spent fuel are (a) an air-cooled, shielded concrete vault, (b) a caisson, consisting of a metal-lined hole in the ground with a concrete plug, and (c) an aboveground concrete cask with natural convection air-flow cooling.

Weights and Volumes

Eventually a choice has to be made between reprocessing the spent fuel, storing it indefinitely in the form as removed from the reactor, or disposing of it by burial or other techniques. Let us consider how much material is involved in the management of spent fuel. The assembly has a uranium weight of around 1000 pounds; other metals add some 200 pounds. The volume of one assembly is 6.7 ft^3, so the 60 assemblies removed per year from each reactor give a total of around 400 ft^3 of spent fuel. From 100 reactors in the U.S., the annual spent fuel volume would thus be 40,000 ft^3, corresponding to less than 1 foot depth in a standard football field (300 ft by 160 ft).

The actual amount of radioactive material in the spent fuel is considerably smaller. From the previous chapter we saw that fission products have a weight of 3.5 percent of the spent fuel. If the fuel were reprocessed, these would be extracted, as would most of the plutonium isotopes. The uranium would be cleaned up in preparation for reuse. For each 1200-pound assembly there would be only around 35 pounds of fission product waste. For 60 assemblies discharged per year this is 2100 pounds, or about one ton. Assuming that the weight of these fission product elements on the average is ten times that of water (which weighs 62.4 pounds per cubic foot), their actual volume would be only 3.4 ft^3, which is 18 inches on a side. This figure is the source of the claim that a year's wastes from a reactor could fit under an office desk. The statement is misleading because most wastes would not be stored or disposed of in such a concentrated form because its radioactivity would produce intense heat. The wastes would be diluted by some solid before being discarded. This simple sketch compares

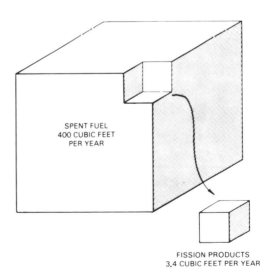

SPENT FUEL
400 CUBIC FEET
PER YEAR

FISSION PRODUCTS
3.4 CUBIC FEET PER YEAR

Comparison of the volume of fission products with the volume of spent fuel. The numbers refer to the production by one reactor in a year.

volumes of fission products and spent fuel produced by one reactor in a year.

Choices of what to do with spent fuel are: (a) store it indefinitely in the form as removed from the reactor, (b) reprocess to extract fission products and recycle other materials, or (c) dispose by burial or other isolation technique. In the next chapter we examine these choices further.

CHAPTER 14

Reprocessing, Recycling, and Resources

Between the mining of uranium and the final disposal of waste products are many processes; these comprise the nuclear fuel cycle. The term "front end" of the cycle refers to preparation of uranium for use in power reactors; the term "back end" refers to operations performed on spent fuel.

The "Once-Through" Cycle

The sketch outlines the steps in the "once-through" cycle for nuclear reactors as it has been operating in the nuclear industry. First, exploration for new uranium deposits is done. Then uranium is mined in several western states of the U.S.—mainly Colorado, Utah, and Wyoming. Natural uranium (0.7% ^{235}U, 99.3% ^{238}U) is extracted from the low-grade ores by a milling process, leaving a large waste residue called mill "tailings."

A complex uranium compound called "yellow cake" is the useful product purified by chemical refining. The compound is then converted to uranium hexafluoride, UF_6, a gas at ordinary temperatures and pressures. UF_6 is shipped to the isotope separation plant, where a process of gaseous diffusion separates an enriched or product stream of ^{235}U content greater than that of natural uranium from a "tails" stream of depleted uranium. The slightly enriched uranium (about 3% ^{235}U),

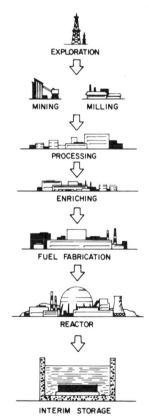

EXPLORATION

MINING   MILLING

PROCESSING

ENRICHING

FUEL FABRICATION

REACTOR

INTERIM STORAGE

The nuclear fuel cycle, carrying uranium from the mine to spent fuel storage in water pools.

still as gaseous UF_6, goes to the fuel fabrication plant, where it is converted to the form of solid pellets of uranium dioxide, UO_2. The pellets are inserted into tubes to form fuel rods, which are sealed, tested, and assembled into bundles for shipment to the reactor site. Each fuel assembly remains in the converter reactor for about three years, after which it is removed and placed in a storage pool for radioactive "cooling."

There are now two choices of what to do with the spent fuel. The first is to continue interim storage in water-filled pools in a "once-through" cycle. The fuel assemblies would then be stored or disposed of intact or in bundles of fuel rods in a waste repository. The second choice is to recycle uranium and plutonium. This is possible through reprocessing the spent fuel by methods similar to those used at Hanford during World War II. Plutonium can be blended with slightly enriched uranium to form a fuel called "mixed oxide" (MOX), which is a combination of UO_2 and PuO_2. The uranium in spent fuel has a ^{235}U content higher than that of natural uranium. It can be returned to the isotope separation plant for reenrichment and reuse. The diagram shows the once-through and recycle modes of operating the nuclear fuel cycle, including points at which wastes arise.

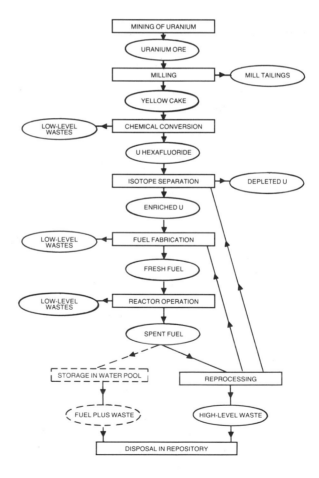

Wastes in the nuclear fuel cycle. Two choices are shown—the once-through cycle and a recycle of uranium and plutonium.

Reprocessing of Spent Fuel

The long-range plans of the nuclear industry of the 1970s included reprocessing of spent fuel.* Thus, when the activity in the fuel stored in the water pool had dropped sufficiently, the fuel would be placed in casks for shipment to a centralized reprocessing plant. Some commercial reprocessing had been done by Nuclear Fuel Services at West Valley, New York. A new reprocessing facility designated as AGNS was built and given testing with nonradioactive chemicals. There it was planned to cut fuel assemblies into pieces, dissolve the contents of the fuel tubes with nitric acid, and apply a PUREX solvent extraction method. This would separate the three principal components uranium, plutonium, and fission products. The flow of materials is seen in the diagram.

There are several advantages of reprocessing, in contrast to a "once-through" use of nuclear materials. The valuable energy resources contained in fissile plutonium-239 and plutonium-241 are made available. By recycling uranium and plutonium the amount of mining and milling of new uranium ore is reduced and the uranium reserves are conserved. The waste volumes to dispose of are smaller for the fission products separated in reprocessing than for spent fuel. Also the removal of plutonium and its subsequent consumption in a reactor eliminates it as a source of hazard in waste disposal. There is a good indication that rare strategic metals could be extracted from the fission products. Disadvantages include the extra cost of the reprocessing facility and its operation, the radiation exposure to workers at the facility, and the increased accessibility of plutonium, which conceivably could be diverted and used by terrorists.

Reprocessing was banned as a U.S. national policy by President Carter in 1977 because it makes plutonium more readily accessible. It was hoped that countries not yet having nuclear weapons would thus be discouraged from developing reprocessing facilities to obtain plutonium. President Reagan in 1981 rescinded the ban and urged the nuclear industry to resume commercial reprocessing. Financial problems and uncertainty about future government policy, however, have discouraged industry from acting.

Other countries such as France and Japan have successfully implemented reprocessing. For example, the company COGEMA in France operates the plant UP3. Started in 1990, it reprocesses 800 tonnes of spent fuel annually. It uses an advanced PUREX process with special features: a continuous nitric acid dissolver for the uranium oxide; noncorrosive zirconium vessels; and a large-capacity unit to mix fission products with powdered glass and melt them together for

---

*For an excellent discussion of reprocessing, we recommend the article by William P. Bebbington, "The Reprocessing of Nuclear Fuels," *Scientific American*, December 1976, p. 30.

eventual disposal as a solid form. The French believe that reprocessing is not more expensive than direct disposal in the long run, that recycling plutonium is very desirable, and that it minimizes risk.

The Breeder Reactor

The fissile content of spent fuel is equivalent to uranium enriched to 1.43 percent in ^{235}U, which is twice the fissile content of natural uranium. More than half the atoms are plutonium, which can be used as starting fuel for "breeder" reactors. The breeder is a reactor that does not have a neutron moderator, and thus fission is initiated by fast neutrons. When fueled by plutonium with the core surrounded by ^{238}U as a "blanket," the breeder produces more fissile material than it burns. In 10 to 20 years it can create a new core of plutonium. The great virtue of the breeder is that it makes use of the more abundant isotope of ^{238}U rather than depending on the less abundant ^{235}U as in the converter reactor. In the once-through mode, the latter uses only half of the ^{235}U in natural uranium. The available uranium supply would last around 50 times longer if the breeder cycle were adopted.

Research and development on the breeder reactor has been in progress since the early 1950s in several countries. Leaders have been the U.S., the U.S.S.R., France, the U.K., and Japan. The U.S. started construction but later abandoned a demonstration breeder called the Clinch River Breeder Reactor Project. The French sodium-cooled fast breeder SuperPhenix was operated for a while and then was shut down because of leaks. Its restart is problematic for political reasons. The European Fast Reactor (EFR) is a breeder being designed as a cooperative effort of France, Germany, and Great Britain. It will be several years before the design is completed. Japan continues to favor reprocessing and breeder reactors and may be the first to successfully revive the breeding process.

Decisions about the breeder affect the management of nuclear wastes. Adoption of breeders automatically requires reprocessing; disposal involves fission products rather than spent fuel assemblies.

The desirability of pursuing research, development, and testing of breeder reactors is a subject of debate. Some believe that the breeder is undesirable because of the production of plutonium, which is both toxic and usable for nuclear weapons. Others, however, note that the safest place to put plutonium is in a reactor, which is itself a very strong source of radiation. As breeder reactors are built and go into operation, they may be fueled initially with fissile plutonium from light-water reactors. Alternatively, they could be fueled with plutonium removed from weapons dismantled as a consequence of disarmament treaties. For fertile material, that serving as the source of new plutonium, breeder reactors can use the large stockpile of depleted uranium from uranium isotope separation over many years.

This has less ^{235}U (about 0.2%) than natural uranium (0.7%), but only the ^{238}U is needed to produce new plutonium in the breeder.

Breeder reactors are likely to remain in the research stage in the U.S. for a number of years. The breeder's ultimate economic advantage is felt only when fossil fuels are in short supply and when natural uranium stockpiles and reserves become too expensive to exploit. Currently, uranium fuel prices are low and form a rather small part of the cost of nuclear electricity. It will be well into the 21st century before breeder reactors are adopted widely in the U.S. On the other hand, breeders may become popular in several foreign countries that do not have large reserves of either fossil fuels or uranium.

Partitioning and Transmutation

A process called "partitioning" separates chemicals as an extension of reprocessing. It extracts, isolates, and irradiates certain long-lived transuranic elements such as neptunium, americium, and curium, and fission products such as iodine-129 (1.7 x 10^6 yr) and technetium-99 (2.13 x 10^5 yr). Partitioning is being considered in order to improve the safety of disposal of high-level waste. In a process called "transmutation" some of the nuclides would be fissioned to become fission products with generally shorter half-life than the parent nuclei. Others would become stable or have shorter half-lives. The phrase "actinide burning" (actinides are elements above 89 in the periodic table) is often used interchangeably with "transmutation."

Two quite different alternative systems are being contemplated in the U.S. The first is a complete system consisting of a metal fuel fabrication unit on site, a reactor operating with fast neutrons, and a chemical extraction process based on melting. The Integral Fast Reactor (IFR) concept would combine the three steps. Neutrons in the reactor would bombard and transmute the transuranic and fission product isotopes contained in the recycled fuel. The second approach would involve a linear accelerator to bring protons up to an energy of 1600 MeV. A current of around 0.1 ampere would strike a subcritical assembly of fuel consisting of transuranics, including plutonium. A high-energy nuclear process called spallation releases as many as 50 fast neutrons per proton. These neutrons are multiplied by factors of 10 to 20 in the target. The resultant neutrons bombard the fuel isotopes, transmuting them into less hazardous forms. In principle, one proton can destroy several hundred transuranic atoms, so that one accelerator could process the wastes from over half the reactors in the U.S.

Although these concepts look promising, several technical, economic, and societal issues need to be resolved, as discussed in the references for this chapter.

We now have a picture of the present nuclear fuel cycle and possible variations on it involving recycling of uranium or plutonium. As was seen in the diagram, some radioactive wastes are produced in every stage. The disposal of these in an environmentally sound way will be discussed in subsequent chapters.

CHAPTER 15

Uranium Mill Tailings

The amount of uranium in ore is quite small, about 0.2 percent by weight in the U.S. Thus there is a large residue called "tailings" from the chemical processing of ore, called milling. Each 1000-megawatt electric nuclear reactor requires about 150 tons per year of natural uranium. The total ore required is 500 times this figure. The map shows the main areas of uranium deposits and mining districts of the United States. Few of these are active because the demand for uranium did not grow as expected and foreign supplies expanded. Between 1979 and 1990 employment (in person-years) in the

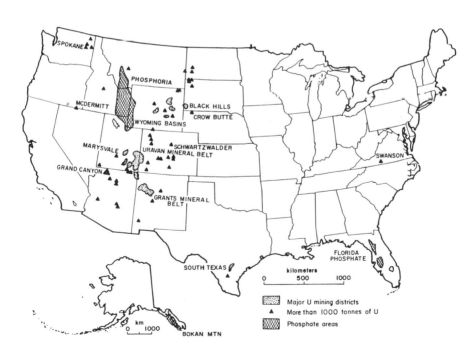

Uranium deposits and mining districts in the U.S. (From *Uranium: 1991 Resources, Production and Demand*, OECD/NEA and IAEA, Paris, 1992.)

uranium industry dropped from 21,950 to 1,340. Natural uranium stocks amount to over five years of operation, and U.S. resources are ample for the foreseeable future.

The tailings contain elements such as thorium and radium, which are mainly by-products of the decay of ^{238}U as was shown in the table on p. 14. Most of these elements are not removed in the extraction of uranium. Generally, tailings leave the mill as a liquid sludge and are allowed to dry. They are collected in piles within enclosures. The photograph here shows a typical tailings pile. Precaution is necessary to prevent tailings from contaminating groundwater or getting into the air as dust. For their radioactive content tailings are no more dangerous than the ore that was removed from the earth, but since the material has been brought to the surface and been converted to a new form, it can pose a hazard unless protective layers of earth are applied.

A mill tailings pile, resulting from the first processing of uranium ore. (Courtesy of U.S. Nuclear Regulatory Commission.)

Radioactivity in Tailings

The tailings are classified as radioactive waste separately from the low-level wastes from power production and from isotope uses. On a short-term basis, the radiation hazard to the general public in the surrounding region from tailings is relatively small. The possibility of erosion by wind and water plus geologic effects requires, however, that the residues be properly stabilized.

One isotope of main concern is radon-222. As we saw on p. 14, it is the daughter of radium-226, which has a half-life of 1599 years.

Radon is produced continuously by the decay of radium left with the ore residue. Radon-222 has a short half-life, 3.82 days, and decays in a chain with four short-lived products. The element radon is an inert gas of the same type as helium and neon. Thus much of it can come out through the pores of the tailings and the soil into the atmosphere and be carried some distance by wind before it decays. The daughters of radon can deposit on surfaces as contamination or can deposit in lung tissue of animals and human beings.

Over the years some 140 million tons of tailings have accumulated, and several million tons more are added each year. Earlier there was little concern about the potential hazard. Since almost all earth and rock contains uranium and thus radium, the tailings were treated essentially as just another form of ore residue. Regulations set by the states were not very rigorous and their enforcement was often lax. Concern about mill tailings arose as the result of use of the sand-like material for fill in building construction in Grand Junction, Colorado. In 1966 it was realized that occupants were being exposed. Mill owners were requested to take better care of tailings. Congress in 1972 authorized funds for cleanup in Colorado, and in 1978 passed the Uranium Mill Tailings Control Act, calling for the Department of Energy to take remedial action at inactive milling sites. A DOE program involving 24 sites in ten states is in progress.

The passage of the National Environmental Policy Act (NEPA) resulted in a new look at the situation and prompted the Nuclear Regulatory Commission to initiate a generic environmental impact statement (EIS) on mill tailings. The EIS has some specific guidelines about ways to handle tailings. It recommends that tailings be located remote from population centers, preferably underground, with a cover 3 meters thick to keep the radon radioactivity that escapes below 2 picocuries per square meter per second. It also recommends that the bottom and walls of a tailings pit be covered with a thick layer of clay and that vegetation be grown above the pit to resist erosion. Extensive studies of ways to control migration of the radioactive material have been conducted. Emphasis has been placed on preventing flooding erosion of slopes by the use of riprap (rock armoring).

Some studies show that the potential hazard to the public from mill tailings can be larger than that from other places in the nuclear fuel cycle, including reactors. The exact degree of hazard from radon and its daughters is difficult to specify, however, because there seems to be little or no correlation between lung cancer incidence and geographical areas having a high level of radon. The trend, however, is toward use of the latest recommendations of scientific committees in planning for control of exposure. The release of radon is not unique to uranium mining since the mining of phosphates for fertilizers brings up radon in comparable amounts. Much additional information and insight on mill tailings is found in the book *Environmental Radioactivity* by Eisenbud. (See references for Chapter 4.)

Indoor Radon

The radiation exposure from radon was quoted in Chapter 5 as 55 percent of the total exposure for the average American. This figure is not the same for all people because some homes and buildings contain more radon than others. Concern about excessive amounts has been growing recently. As early as 1906, the famous scientist Lord Rayleigh is said to have known of radioactivity in granite, and geologists over the years had studied the uranium content in rocks and water. Attention was called to the problem of radon in homes by Henry Hurwitz, Jr., who noted that federal standards for emissions from nuclear power plants were much more stringent than those on radon. Radiation measurements made in the aftermath of TMI-2 led to the discovery of high radon concentrations in Reading Prong, a geological formation mainly in Pennsylvania. There some homes had concentrations hundreds of times the average.

There is some uncertainty as to the actual hazard, which is estimated in part from cancer cases among uranium miners. These workers have historically been at risk, especially since they tend to be heavy smokers as well. It is difficult to separate the effects of radon and smoking. Also the work environment of miners is quite different from that of a typical person. Another factor is the difficulty in modeling the dose-effect relationship.

Although the hazard is commonly associated with radon, it is the short-lived daughter products—isotopes of polonium, bismuth, and lead—that are the principal offenders. One might think the dose would be due to the radioactivity of decay products arising from inhaled radon. In fact, it is largely the result of breathing air that contains radioactive daughters attached to dust particles along with unattached decay products. The amount of radiation received thus depends on the relative amounts of such materials the body traps or eliminates. Further discussion of the modeling of radon effects, including the use of the Working Level unit of exposure, is found in the book by Shapiro (see References).

For many years the contribution of radon to the average public radiation exposure was ignored, but in 1987 the NCRP included it in published estimates. The Environmental Protection Agency mounted a public information program, issuing in 1986 and updating in 1992 the bulletin *A Citizen's Guide to Radon*. The booklet explains the origin and nature of radon and recommends testing of individual homes and corrective action if the radon concentration in air is above 4 picocuries per liter (pCi/l). The publication gives telephone numbers of radon offices in all the U.S. and territories. The EPA estimates that about 14,000 lung cancers appear each year due to radon, based on the linear dose-effect relationship, similar to the one that is used for other radiation hazards. Putting the risk on an individual basis, at the 4 pCi/l level a non-smoker has an increased risk of about one chance in 1000 of developing lung cancer, while a smoker has one chance in 100.

Radon in buildings comes primarily from the ground, which naturally contains a varying amount of uranium. The amount of radon present in the air at any time depends on the degree of ventilation because outside air has a low concentration of around 0.1 pCi/l. The sealing of rooms, however, to conserve energy does not account for the large amounts of radon found in some locations. It appears that the radon is "pumped" through cracks in basement floors or walls by differences in pressure, temperature, and wind velocity.

Methods of control include thorough sealing of cracks and the installation of pipes and fans that will exhaust the gas from under floors to the atmosphere and prevent its getting into the main living areas. The expense of such modifications varies, but is estimated to average around $1000 per building. To determine whether action is needed, the EPA recommends that one or perhaps two short-term measurements be made, using commercially available kits costing around $25. Since radon measurement results can fluctuate greatly from time to time, a test lasting several months is in order before incurring a large expense. It is considered better to include radon protection in the original construction of a house than to treat it later. Some researchers (see References) believe that a concerted effort should be made by the EPA to find and correct levels in excess of 20 pCi/l around the country, thus protecting those who are in greatest hazard.

Appreciation is extended to Dr. James E. Watson, Jr., for his thoughts on the radon problem.

CHAPTER 16

Generation and Treatment of Low-Level Waste

How Low-Level Wastes are Defined

Radioactive wastes are classed officially as low-level wastes (LLW) if they are not spent fuel, high-level reprocessed wastes, mill tailings, or transuranic wastes. In general, LLW contain a small amount of radioactive material in a rather large volume. Thus they usually do not require shielding or heat-removing equipment, and most are acceptable for near-surface land disposal. They are mainly by-products of the operation of nuclear reactors and the use of radioisotopes. Although some LLW come from DOE and defense programs, our present focus is on commercial wastes, both those from the nuclear fuel cycle and those from the "non-fuel-cycle," which will include industrial and institutional wastes. We shall discuss also two special types of LLW, coming from decontamination and decommissioning.

Low-Level Wastes from Nuclear Power Plants

Of the low-level radioactive wastes generated today, more than half of the volume and most of the activity comes from nuclear power plants. The radioactivity arises from two nuclear processes: activation and fission.

Activation results from the absorption by nuclei of neutrons in the nuclear reactor. Targets are metals in the core and vessel, corrosion products in the cooling water or on surfaces, or impurities in the water. Since much of the internal structure of a reactor is composed of stainless steel, isotopes of iron, cobalt, nickel, and manganese are produced. An example is cobalt-60, half-life 5.27 years, emitting gamma rays of 1.25 MeV average. It is formed by slow neutron reaction with cobalt-59. The table lists data on the important activation products. A certain amount of tritium (hydrogen-3), with half-life 12.3 years, is present as a result of occasional fission into three particles. Tritium also is found in the cooling water of reactors that use boron as a soluble control absorber.

Fission products serve as the second large source of low-level radioactivity. Although most of these are retained within the fuel tubes, a

Activation Products in Reactor Coolant

Isotope	Half-life, yr	Radiation Emitted	Parent Isotope
Carbon-14	5715	Beta	Nitrogen-14*
Iron-55	2.73	x-ray	Iron-54
Cobalt-60	5.271	Beta, gamma	Cobalt-59
Nickel-59	7.6×10^4	x-ray	Nickel-58
Nickel-63	100	Beta	Nickel-62
Niobium-94	2.4×10^4	Beta, gamma	Niobium-93
Technetium-99	2.13×10^5	Beta	Molybdenum-98 Molybdenum-99†

*(n,p) reaction
†Beta decay

few appear in the cooling water. Some are from neutron irradiation of small amounts of uranium left on the outside of fuel rods during fabrication or residual impurity of natural uranium. Additional fission products come from pinhole corrosion leaks in the fuel tube. Because it is impractical to shut the reactor down to remove an offending fuel assembly, some leakage into the water is permitted. Some of the most important beta-gamma-emitting fission products are iodine-131 (8 day), cerium-144 (285 day), cesium-134 (2.1 yr), cesium-137 (30 yr), technetium-99 (2.13×10^5 yr), and iodine-129 (1.7×10^7 yr). Many other chemical species, however, are represented in the coolant, including carbon-14 (5715 yr) resulting from neutron bombardment of nitrogen in air.

Volume Reduction Techniques

The cooling systems of reactors are closed loops, but leaks of water occur, and radioactive components must be repaired or replaced, resulting in the spread of contamination. Also, the coolant is treated to remove impurities, giving rise in time to significant volumes of radioactive material. Because of increasing costs of shipping and disposal of LLW, nuclear utilities have expanded their onsite processing facilities.

Waste management programs include preventing contamination, segregating wastes by type, reducing volume, and conditioning of materials in preparation for shipment and disposal. Many of the practices are dictated by compliance with federal regulations on waste disposal. The principal methods, applied to low-level wastes both at nuclear power plants and defense waste sites, are as follows.

Filtration is much like the process of straining grounds in a coffee pot. The mixture of solid particles and water is forced through a medium that is porous to liquid, as shown in the diagram. When the filter cartridges accumulate enough material, they are removed and packed in a drum for disposal. The 55-gallon steel drum is very popular because of its low cost and convenient size.

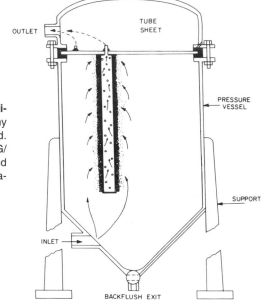

Filter for removing solid particles from water. One of many filter cartridges is sketched. (Adapted from report NUREG/CR-0142 by A. H. Kibbey and H. W. Godbee, Oak Ridge National Laboratory.)

Ion exchange is used if the waste is in solution and cannot be filtered. The process removes contaminants of both the activation and fission types. Charged atoms in the liquid replace other nonradioactive atoms on the solid surface of a resin bed. Eventually the resin becomes saturated and must be either washed clean and regenerated for future use by treatment with a chemical or disposed of as waste.

Evaporation is simply boiling to remove water as steam, leaving the waste more concentrated. A residue of sludge collects in the bottom of the vessel shown in the diagram. The condensate water may be recycled for other uses while the sludge is removed periodically.

Incineration is used for a variety of combustible materials. Ordinary materials such as paper, cloth, wood, and plastic become slightly contaminated and can be burned, reducing volume by a factor of 20 or more. The incinerator shown is of Japanese make. The air containing the smoke from burning passes through filters to remove particles. The gases coming from the process thus contain very little radioactivity and can be released safely with dilution to the atmosphere. The ashes are radioactive and must be treated as low-level waste.

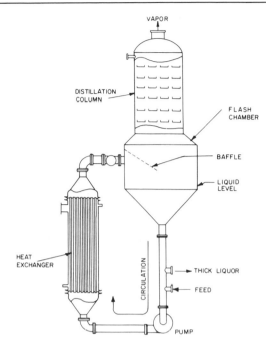

VAPOR

DISTILLATION COLUMN

FLASH CHAMBER

BAFFLE

LIQUID LEVEL

HEAT EXCHANGER

CIRCULATION

THICK LIQUOR

FEED

PUMP

Evaporator for reducing waste volume. The product is a concentrated sludge removed from the bottom of the vessel. (Adapted from report NUREG/CR-0142 by A. H. Kibbey and H. W. Godbee, Oak Ridge National Laboratory.)

Compaction is used to reduce the volume of waste and also to produce a more stable structure for disposal. Different degrees of compaction are possible, depending on the force exerted by the compactor and the nature of the material. The typical compactor has a plate driven by a hydraulic press. The one shown can either compact the contents of a 55 gallon drum or crush the drum and contents. Volume reduction factors of three are possible. A special compactor called the supercompactor can be used to apply pressures up to thousands of pounds per square inch.

Solidification means mixing wastes with some solid material that will resist attack by water after disposal. One popular material is concrete, in which the water remaining in the low-level waste is taken up by the concrete as it sets. The diagram shows a portable cement supply and a concrete mixer used at a nuclear power plant. Another medium is

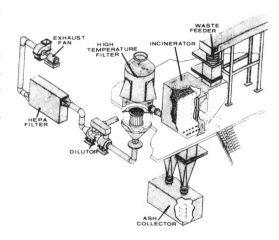

EXHAUST FAN

HIGH TEMPERATURE FILTER

WASTE FEEDER

INCINERATOR

HEPA FILTER

DILUTOR

ASH COLLECTOR

Incinerator for combustible materials. Radioactivity concentrates in the ashes, and the gaseous effluent is diluted to safe levels. (Courtesy of Hitachi Zosen Corp. of Japan.)

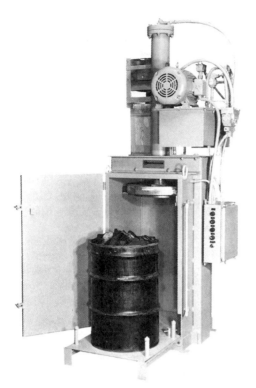

Compactor for waste drums. The plate can either compress the contents of a 55-gallon drum or crush a drum full of waste. (Courtesy of Consolidated Baling Machine Co., Brooklyn, New York.)

asphalt. The diagram (on the next page) shows how reactor wastes and asphalt are fed into a heated grinder that pulverizes, removes water, and mixes. The viscous product pours into a 55-gallon drum, where it solidifies. Another material is polyethlyene, which is easy to handle, is relatively light, and has favorable resistance to water. Solidification stabilizes the waste physically and chemically, making it less subject to leaching when the material is buried.

Another way to protect wastes from chemical attack is to use an officially designated "high-integrity container" made of plastic or concrete and reinforced for stability. It is approved by the NRC as able to last for 300 years under typical exposure to radiation from the contents.

The choice of volume reduction process and equipment depends on the types of wastes and their volumes, on economic factors, and on regulatory requirements. For example, an incinerator is of no use in handling failed contaminated metal equipment; a compactor will not

Solidification using concrete. Water needed to produce concrete comes from the wastes. (Courtesy of LN Technologies, Columbia, South Carolina.)

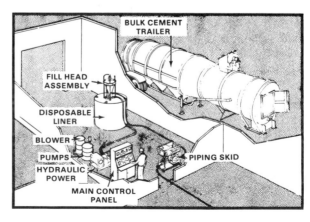

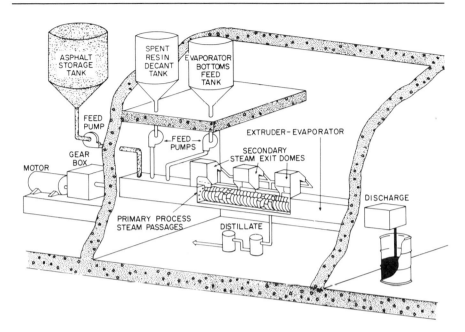

Solidification using asphalt. The heated grinder dries waste, and the molten waste-asphalt mixture solidifies in 55-gallon drums. (Courtesy of Werner-Pfleiderer Corp.)

compress sludges; a million-dollar supercompactor is not needed to handle a few cubic meters per year.

A number of nuclear power plants have installed full waste treatment systems designed to achieve optimum technical and economic performance. A typical example is the Surry Radwaste Facility of Virginia Power. The computerized facility is located in a five-story building. As described by Linwood Morris and William Halverson,* the facility collects waste liquid, separates oil and solids, and evaporates water using heat recovery. Concentrated waste is thoroughly mixed with bitumen and solidified for disposal. Effluent water has minimum radioactivity. The facility also has a shredder-compactor and a hot machine shop.

Low-activity fuel cycle wastes also come from the uranium chemical conversion process and from fuel fabrication plants. The volume of these is less than one-tenth that from the nuclear power plants that they supply.

---

*"Design and Operation of the Surry Radwaste Facility," American Society of Mechanical Engineers conference in Prague, September 1993.

Other Sources of Low-Level Wastes

Non-fuel-cycle wastes are often called industrial and institutional wastes. They are generated by industrial organizations that supply or use radioactive materials, e.g., pharmaceutical companies, and by institutions such as clinics, hospitals, medical research laboratories, and universities. Medical wastes include animal carcasses and other biological waste, trash, various liquids, sealed radiation sources, and technetium-99m generators. Industrial and institutional wastes stem from many uses of many different radionuclides. Larger generators use some of the same treatment methods employed by nuclear power plants.

One special type of liquid waste is scintillation detector fluid used as tracers in biomedical research and medical tests, and for measurements of radiation. Millions of vials of these solutions, used for counting radioactivity, are disposed of each year. They typically contain the chemicals toluene or xylene, with a slight content of tritium or carbon-14. The beta decay of these isotopes triggers the release of light that is detected by a sensor. The total annual activity in scintillation vials in the U.S. is small, only about 10 curies per year, but because of their chemicals they often pose a disposal problem. In some cases the fluids have a low enough activity that they can be disposed of in a sanitary sewer. Disposal sites will not accept such liquids for burial. Incineration is regarded as the best disposal method.

The radioisotope most widely used as a tracer for medical diagnosis is technetium-99m, half-life 6 hours. This gamma emitter is extracted from a longer-lived isotope molybdenum-99, half-life 66 hours. The ^{99m}Tc is said to be "milked" from the ^{99}Mo "cow." The half-life of technetium is so short that holding any residues or contaminated material for decay is preferable to shipment to a disposal site. Reduction by a factor of more than a million occurs in 20 half-lives, which is only 5 days for this isotope. Rather than dispose of the generator (as the ^{99}Mo source is called) when it has weakened, it is preferable to send it back to the supplier, who will combine material to produce a new generator.

Several medical isotopes have half-lives of only a few days; these can be held for decay and thus pose no disposal problem. Some institutions have tended, however, to ship such wastes away for disposal to avoid the need for monitoring and surveillance. Only iodine-125 (59.4 days) has too long a half-life for convenient long-term storage. In bioresearch, however, the most important and widely used isotopes are tritium (12.3 yr) and carbon-14 (5715 yr), which must be disposed of by other means.

Although it is generally assumed that all radioactivity should be avoided and controlled, there are amounts so low that they can safely be ignored. These are treated as exempt by the NRC in the *Code of Federal Regulations,* Parts 30.15 and 20.2005. Examples are 25 mCi of tritium and 100 µCi of promethium-147 as used in self-luminous watches. Also scintillation detector fluids or animal carcasses may be disposed of as ordinary biological wastes if the tritium or carbon-14 content is less than 0.05 Ci per gram, with certain limits on annual disposal. Such amounts are termed *de minimis*; this comes from the Latin phrase "*de minimis non curat lex*" and is translated as "the law does not concern itself with trifles." A similar term is "below regulatory concern."

Amounts of Wastes Produced

Knowledge of the characteristics of wastes is important at every stage of management for several reasons—to ensure radiation protection during processing, to achieve proper segregation and packaging, and to meet regulations on shipment and burial. The features that must be known are volume (ft^3 or m^3) according to waste stream, radioactivity as a specific activity (Ci/ft^3 or Ci/m^3) preferably by isotope, and heat generation rate (W/ft^3 or W/m^3). "Source terms" represent such data, which can be used to plan waste management programs.

Wastes shipped from generators to processing or disposal facilities are always accompanied by a manifest, which is a shipping document that lists properties of waste container contents. The generator is responsible for the accuracy of the information, and the disposer makes spot checks to verify. Some of the waste radionuclides have extremely long half-lives and correspondingly small activities, and are difficult to detect. Examples are technetium-99 (2.13 x 10^5 yr) and iodine-129 (1.7 x 10^7 yr). Predictions of amounts are made using "scaling factors" against readily measured fission products like 30-year cesium-137. Detailed studies of actual amounts of the long-lived nuclides in waste are made to establish these factors. This practice assumes that waste streams are rather consistent.

The chart shows some calculated data on the typical annual amounts of different LLW streams produced by a nuclear plant of PWR and BWR types. The data were developed by the Oak Ridge National Laboratory for the Department of Energy. Since the numbers refer to 1000 MWe power, they are roughly the annual waste from one reactor. We see that both the volume and the activity generated by a BWR exceeds that for a PWR.

In the table that follows, volumes, activities, and concentrations for LLW are compared for DOE sites and commercial sources, including industrial and institutional. We note that the amounts of academic and medical low-level wastes are relatively small compared to those from the generation of nuclear power. The sources of these wastes, radioisotopes, are, however, vitally important to the individual users. Dr. G. John Weir, Jr., of the American College of Physicians, notes (see References) "The cost of LLRW waste disposal and fear of adverse public reaction is causing universities and other institutions to

**Volume and Activity of LLW by Waste Type and Reactor Type,
per 1000 MWe power**
(Alan Icenhour and Steve Loghry of Oak Ridge National Laboratory)

Waste Type*	BWR		PWR	
	m^3/yr	Ci/yr	m^3/yr	Ci/yr
Wet	229	2,330	81.6	440
Dry	374	19.5	246	17.1
Other	5.72	7.72	9.18	0.491
Routine	609	2,360	337	458
Nonroutine	3.73	29,400	0.876	270
Total	613	31,800	338	728

*Wet: spent resins, concentrates, evaporator bottoms, filter sludges, filter cartridges.
Dry: compressible active waste, contaminated equipment.
Other: sand, building rubble, biological waste, etc.
Nonroutine: irradiated components, control rods, core components.

Low-Level Wastes Added Annually to Disposal Sites
(Adapted from data in DOE/RW-0006, Rev. 9, March 1994)

Source	Volume in 1000s of m^3	Activity	
		1000s of Ci	Ci/m^3
DOE sites	40.30	631	15.66
Commercial			
academic	1.255	1.724	1.37
government	4.479	40.780	9.10
industrial	25.725	100.090	3.89
medical	0.743	0.398	0.54
utility	17.162	857.110	49.94
Total	89.66	1631	18.19

pressure researchers to diminish use of radioactive materials. This will impede research, increase the cost of research and slow development of new pharmaceuticals."

Decontamination

The operation over a period of time of any nuclear installation, be it fuel fabrication plant, nuclear reactor station, or reprocessing facility, involves contamination by radioactive substances. Cleanup during operation is called decontamination, while steps taken after shutdown are called decommissioning.

Decontamination means the removal of radioactive material from surfaces such as building floors or walls, hand tools, and the insides of vessels. The object is to reduce radiation exposure to persons working in the area. The technique should be nondestructive and safe to apply; otherwise personnel might receive more person-rems than the cleanup would save. (For a discussion of current methods in use to reduce personnel exposure, see the paper by Y. B. Katayama, et al., in the references.)

Methods are chosen according to the items to be decontaminated. Small tools and equipment can be hand-scrubbed with cleaning agent and cloth or steel wool by an operator wearing a respirator. Other equipment responds to ultrasonic cleaning, in which a solution is vibrated at a high frequency such as 25,000 cycles per second. Removable reactor components may be cleaned also with acid baths, electrical currents, sandblasting, strong chemicals, water jets, and steam. Some surfaces have strippable coatings. Some inaccessible pipes or vessels may be cleaned by circulating high-temperature chemicals. Large volumes of water can be decontaminated using filters and ion exchangers. The Three Mile Island Unit 2 reactor recovery program provided a major test of decontamination techniques. The program also developed an array of ingenious underwater remotely-operated tools to dislodge and safely remove highly radioactive deposits.

In the future, an increased use of robots will enhance both safety and economy. Some of these electrical/mechanical devices can rapidly sample and analyze the contents of a tank of waste. Others can enter an area, survey it for radiation, and clean up contamination. It is possible to adapt for nuclear use robots that are being developed for military activities, or handling powerful explosives, or hazardous waste operations.

Decommissioning

Decommissioning means the removal of a nuclear power plant or other nuclear facility from service after its useful life and taking the necessary steps to protect the public from residual radiation hazard.

The useful life of a reactor is the time between its start-up and the point when repairs and replacements are excessively expensive or safety would be compromised by continued operation. Most commercial reactors are licensed to operate for 40 years. In view of the high capital cost of new reactors, strong efforts are made to continue operation for at least that long.

The first action in decommissioning any reactor is to remove and dispose of the spent fuel, which automatically reduces the inventory of radioactive materials greatly. Several choices then are available: (a) "mothballing," officially called SAFSTOR, in which some decontamination is done but then the plant is closed and guarded, perhaps indefinitely, (b) entombment, or ENTOMB, in which concrete and steel barriers are erected to seal in radioactive components, (c) dismantlement, or DECON, involving removal and disposal of all radioactive material. A variant is delayed dismantlement, allowing for decay of many of the radioisotopes. Another possibility is to convert to a non-nuclear heat source, retaining as much of the equipment as possible.

Decommissioning requires considerable planning to carry out the task safely and economically. After fuel is removed, the system will be decontaminated to a certain extent. Next will be disassembly using torches and jackhammer while rigorously controlling the spread of contamination. The internal structures of the reactor vessel will contain activation products of the kinds listed earlier. They will be in the form of deposits on surfaces and embedded in the metal. Cobalt-60 dominates for the first 50 years or so, and over the long term nickel-59 (7.6×10^4 yr) and niobium-94 (2.4×10^4 yr) are of concern. Many components will be cut up and buried as low-level waste. Highly radioactive materials will exceed limits for near-surface burial and will have to be dealt with separately. A promising alternative to cutting up reactor vessels for disposal is melting, followed by removal of most of the radioactivity, with the purified metal recast into radioactive waste storage or shipping containers. The residual radiation will be inconsequential.

Cost estimates for decommissioning vary greatly, but a typical figure is $150 million. This is a relatively small fraction of a $2 billion cost of a plant and is very small compared with the $12 billion value of the electricity produced over the life of the plant.

Projections

The amounts of low-level radioactive waste to be disposed of in the future depend on several factors:

(a) The extent of volume reduction by generators. In the 1980s the nuclear utilities greatly reduced the amount of low-level waste by several techniques—avoiding contamination of inert materials, screening to separate radioactive and nonradioactive trash, reducing water leakage, compaction of dry active waste, and incineration. The bar graph shows the dramatic decrease per U.S. power reactor over a span of a dozen years. The purpose of such efforts was to cut costs of waste transportation and burial. In the future, the utility industry can still effect some further reduction of volume and activity, but large additional changes are not likely.

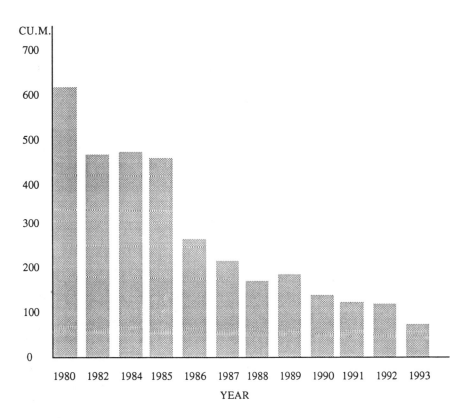

Volume of low-level radioactive waste per U.S. nuclear power reactor, 1980-1993. BWRs and PWRs are represented. The number of cubic meters dropped by a factor of more than eight. (Adapted from *Annual Report 1993,* Institute of Nuclear Power Operations.)

(b)   The role played by waste processors through recycling programs. These can significantly reduce the amount of low-level metal waste that has to be buried. An example is Scientific Ecology Group (SEG) at Oak Ridge, Tennessee, which operates a metal processing facility. Most of the radioactivity is removed from lead, stainless steel, carbon steel, copper, aluminum, and alloys. Useful products are prepared, such as radiation shield blocks, canisters for high-level waste, and reinforcement bars for concrete.

(c)   The development of new technology. Biological wastes are troublesome to store and dispose of. They usually are only slightly radioactive, as the result of medical or research tests. Refrigeration of animal carcasses is possible, but expensive, and treatment with lime and packaging in sealed drums for disposal is conventional. A promising new technique uses hot sodium hydroxide in a simple vessel to dissolve tissue and destroy bone structure. With only slight dilution, a nearly neutral sterile solution can be disposed of in the sewer. Costs are very small compared with burial.

(d)   The number of reactors that are shut down and decommissioned by one mode or another. If the cost of operation, maintenance, and major repair of a power reactor is excessive, it may be shut down before its license expires. If then it is decommissioned at once, a great deal of low-level waste is generated. Plants that have only one reactor are more likely to suffer that fate than those with several reactors on the same site.

(e)   The number of reactors that secure a license extension beyond 40 years. For reactors that have been operating productively, efficiently, and inexpensively, a decision to seek a license extension from the NRC is regarded as wise. The capital cost will have been amortized, and the cost of refurbishing is low compared with that of building a new plant. Some additional low-level waste is generated in the accompanying cleanup operation.

(f)   The rate at which new advanced light-water reactors are developed and put into operation. The timing of their appearance and the number of units built will affect waste generation. Experience, however, with conventional reactors will suggest materials choices and design changes that will allow some reduction in waste generation. Zero-leakage fuel and nonactivating components would be required to achieve the ideal of a reactor that produces no low-level radioactive wastes.

The several factors cited above depend on both local and general economic conditions and on social acceptance, which are difficult to predict even in the short term. The results of a study by the Department of Energy of the U.S. nuclear capacity are shown in the graph. In the "No New Orders" case, all reactors retire when their license expires, and no new ones are built. In the "Upper Reference" new orders meet a growing demand for electric power. The "Lower Reference" involves a lower assumed economic growth and a smaller fraction of power that comes from nuclear. The part of the curve labeled "Common" applies to all cases up to the year 2010.

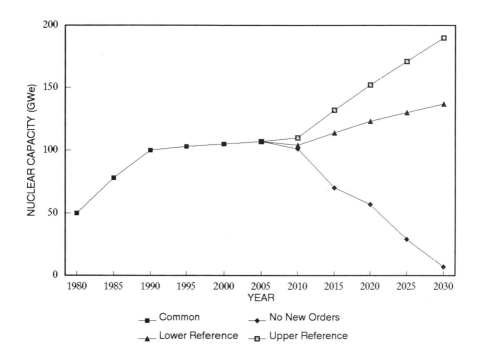

Historical and predicted trends in U.S. nuclear capacity, 1980-2030. For the period after 2005, assumptions are optimistic, realistic, and pessimistic. (Adapted from Department of Energy roport DOE/EIA-0438 (91).)

CHAPTER 17

Transportation of Radioactive Materials

Radioactive materials of many types are transported by public and private carriers all over the country. Many of the shipments are medical isotopes, some are institutional wastes, and many are low-level wastes from nuclear reactor operations. Spent fuel is occasionally transferred between reactor stations. When the federal government begins to accept spent fuel for disposal as high-level waste, the number of shipments will increase significantly.

Concerns About Shipping

People tend to be apprehensive about transport of radioactive materials through or near their communities. Their concerns stem from a basic fear of radiation, awareness of the likelihood of accidents, and inadequate knowledge of the physical protection provided for such materials. People perceive a hazard from radiation emanating from contents of the vehicle as well as from contamination that might result from rupture of a waste container. The public reacts by making one or more statements such as (a) no shipments should be made, (b) shipments should be made to avoid our town, (c) we should know in advance the route to be taken and the schedule of shipment, (d) the vehicles should have a prominent marking "radioactive," (e) wastes should be shipped inconspicuously to avoid hijacking, (f) guards or an escort vehicle should accompany the shipments, (g) a radiological rescue team should be available at all times, and (h) shippers and carriers should be heavily insured.

Many expectations of the public are being met through physical security and adherence to regulations. However, a complete ban on transport is unrealistic because radioactive materials, including wastes, exist and must be moved. It is not possible to transport wastes without coming near areas with some population by the very nature of the ground transportation system. In this chapter we give some accident data and examine the design requirements on packaging of radioactive materials to withstand impact, fire, and immersion.

Facts About Transportation

We can provide some statistics on transportation in general and nuclear materials in particular. Some 500 billion packages of commodities of all kinds are transported in the U.S. each year. Motor carriers handle 57 percent, railroads 38 percent, and waterways about 5 percent. Less than 1 percent goes by air.

Hazardous materials shipments are a very small part (1 in 5000) of all goods and materials transported, and radioactive materials shipments are a rather small part (1 in 50) of the hazardous materials. Therefore, one shipment in 250,000 contains radioactive material. The sketch shows the percentages of shipments, and the tables give recent U.S. data on radioactive material transport according to its origin.

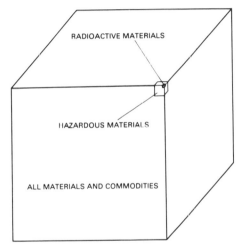

Shipments of goods in the U.S.
Radioactive materials form 1/50 of hazardous materials, which are 1/5000 of all items shipped.

Annual Shipments of Nuclear Materials in the U.S.
(Data courtesy of Sandia National Laboratories.)

Type of Material	Shipments per Year
Exempt amount or limited radioactive level, e.g., smoke detectors, luminous signs or watches	700,000
Pharmaceutical and other medical sources, e.g., radioisotopes used for diagnosis or treatment	910,000
Industrial radiation sources, e.g., gages to measure thickness of paper, portable x-ray devices	220,000
Nuclear materials, e.g., uranium, fresh fuel from fabrication plants, interplant spent fuel	200,000
Wastes from all industrial and medical sources besides nuclear power plants	100,000
Nuclear power plant wastes	50,000
Total	2,180,000

The level of public safety in the transport of radioactive materials can be put in perspective by examining the table. It refers to a year in which a breakdown of data by class of hazardous materials is available. Only 0.4 percent involve radioactivity. It should be noted that the figures refer only to accidents that are reported to the Department of Transportation. Some events of each type fail to be reported.

One Year's Hazardous Materials Incident Reports
(to the Department of Transportation)

Class	Number of Reports	Percent of Total
Flammable liquids	2,347	39.4
Corrosive materials	2,239	37.6
Combustible liquids	381	6.4
Poisons, type B	286	4.8
Miscellaneous	178	3.0
Flammable compressed gases	155	2.6
Oxidizers	137	2.3
Nonflammable compressed gases	125	2.1
Flammable solids	36	0.6
Organic peroxides	36	0.6
Radioactive materials	**24**	**0.4**
Blasting agents	6	0.1
Poisons, type A	6	0.1
Total	5,956	100

The small number of incidents involving radioactivity is borne out by a study conducted by the International Atomic Energy Agency in 1986.* It revealed that fewer data are available than desired, but stated regarding transportation, "Reviews of the available historical data have shown that there has never been a serious incident involving the dispersal of radioactive material."

From time to time, communities or states pass ordinances or laws prohibiting the movement of radioactive materials. Some of these are not enforced. Confrontation between local and federal jurisdiction is sometimes avoided by rerouting. In other cases the federal government, through the Department of Transportation, asserts its authority on the basis that national laws take precedence over local laws. The so-called "commerce clause" stating that restraint of trade is illegal is sometimes invoked.

---

*IAEA-TECDOC-398, Vienna 1986.

Boxes, Cans, and Casks

The philosophy of protecting the public against radiation in transportation has been that accidents will occur and that protection must be provided by the packaging of the radioactive materials. The degree of protection is selected on the basis of the level and type of radioactivity contained. Type A packages (see the diagram of a typical one) can contain a limited number of curies and are designed to withstand normal wear and tear of transport exclusive of accidents. An inner glass bottle is protected by a metal can, a fiberboard insert, a lead container, and a fiberboard box. Even if the package were badly damaged, there would be little hazard because of the small radioactive content. Type B packages (two views of an example container are shown) are designed for a larger number of curies and must withstand accidents without leaking. Important features are the holddown devices, heavy steel wall, fire protection, and internal suspension.

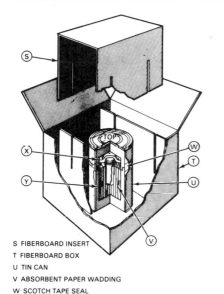

S FIBERBOARD INSERT
T FIBERBOARD BOX
U TIN CAN
V ABSORBENT PAPER WADDING
W SCOTCH TAPE SEAL
X TOP SECTION LEAD CONTAINER
Y BOTTOM SECTION LEAD CONTAINER

Diagram of a representative type A shipping container, which can vary markedly in size and materials. (Courtesy of Department of Transportation.)

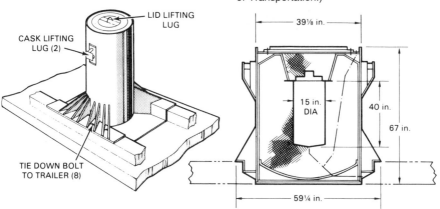

Type B transport cask used by Chem-Nuclear Systems, Inc. The capacity is 4 ft^3; lead shielding is over 10 inches thick; the cask lid weighs 1600 pounds. (Adapted from a Department of Transportation report.)

The shipping cask for spent nuclear fuel is a still more rugged and elaborate container. It provides four types of physical protection:

1. containment, to prevent material from being released into the environment
2. shielding, to prevent radioactive exposure to employees or passersby
3. heat management, to remove the energy released by decaying fission products
4. criticality prevention, to avoid accumulation of enough fissile material to be multiplying.

In the era when reprocessing was anticipated, it was planned that spent fuel would be sent from the nuclear plant to the reprocessing facility after a brief "cooling" period in the water pool. Casks were thus designed with circulating water to remove much of the decay heat. The IF-300 model of General Electric Company is an example of the early design. The two cutaway views show its features. It will hold 7 PWR or 18 BWR assemblies.

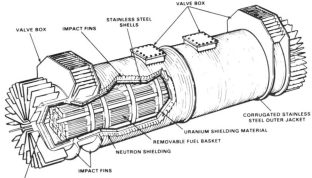

Cutaway view of a spent fuel shipping cask. PWR or BWR fuel assemblies are held in a fuel basket. Water and lead serve as shielding. Fins help remove heat to the air. (Courtesy of Pacific Northwest Laboratory.)

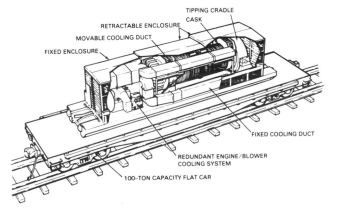

Spent fuel cask in normal arrangement for shipment by railroad. (Courtesy of General Electric Company.)

More modern versions of spent fuel casks do not contain water because the fuel has been stored for a long time and decay has reduced its heat-generation rate. Also, the shipping casks may double as storage containers or as disposal canisters, requiring additional features.

Cask for storing and shipping spent fuel. (Courtesy of EG&G Idaho, Inc.)

Fuel casks are built with massive amounts of fracture-resistant metal such as ductile cast iron or stainless steel. For added protection, large end fittings called impact limiters are attached. These are made of balsa wood or cork, which will absorb much of the energy in any impact.

There are very few casks available in the U.S. for the movement of spent fuel because most of it has been stored where it was produced. Several of those in use are old and outmoded in terms of new standards. Looking forward to 1998 when the Department of Energy starts taking title to commercial spent fuel, a vigorous effort will be required to get enough NRC-licensed casks to meet the demand. Many companies in the U.S. and abroad would like the business, but they can contribute only if there are orders.

Such casks, like all of those containing enriched uranium or plutonium, must be safe from nuclear criticality. As in spent fuel pools, protection is provided by separation and use of neutron absorbers. A combination of experiments and computations provides the design features for spent fuel casks. Computer codes such as ORIGEN determine the concentrations of fuel and absorber isotopes, while statistical (Monte Carlo) codes such as KENO V in SCALE are used to estimate neutron multiplication factors. Such information permits the designer to apply "burnup credit," which is the reduction in multiplication in fuel resulting from the formation of absorbing fission products. Typically a drop of 20 percent occurs in fuel that has been exposed to 20,000 megawatt-days per metric ton (1000 kg).

For transporting transuranic wastes (TRU) to the Department of Energy's WIPP site in New Mexico (see Chapter 20), a B-type container called TRUPACT II has been built. It is designed for use with "contact-handled" TRU, which is 98 percent of the total. The cask is cylindrical, with a flat bottom and domed lid. It holds two layers of seven 55-gallon drums of waste. The containers, with a capacity of 3290 kilograms, are doubly contained and sealed. They can be carried by truck or rail. TRUPACT II passed the required NRC accident tests, and has been in regular use for several years without accident.

Transfer of plutonium from France to Japan by freighter brought protest from persons concerned about a possible attack of the vessel, or hijacking by terrorists, or an accident that would contaminate the sea. A novel suggestion has been made—to use submarines for this purpose.

Transportation Safety

Wastes are but a small part of the radioactive materials transported, but the same regulations apply to all. On a world basis, the applicable document is the International Atomic Energy Agency's "Regulations for the Safe Transport of Radioactive Material," first issued in 1959 and frequently updated. These rules are compatible with those of the U.S. appearing in the Nuclear Regulatory Commission's 10 CFR 71 and the Department of Transportation's 49 CFR 171 and following sections. Such rules derive from recommendations by standards-setting bodies such as ICRP (see Chapter 6), which apply fundamental radiological protection principles.

Regulations are based on concepts such as ALI (Chapter 6). Numerical values are prescribed for the maximum contents of packages in both curies and curies per gram, according to type of container, the material form (concentrated or dispersed) and the particular isotopes involved. For example, the maximum activity for a sealed californium-252 source is two curies; that for normal tritium is 1000 Ci; that for uranium is unlimited, but the number of curies per gram allowed depends on the enrichment.

For many years, shipping casks have been designed to meet stringent rules of the Nuclear Regulatory Commission. A cask must be able to withstand a sequence of events as damaging or more damaging than those that would actually be encountered. The first test is a 30-foot fall on a flat, unyielding surface. Such a test resembles a fall from an overpass onto concrete. However, it should be noted that all real surfaces, such as ground or highways, are somewhat yielding, so that the design requirement corresponds to a drop from a much greater height. The second test is a 40-inch fall onto a metal pin 6 inches in diameter. This is similar to a sharp corner of a bridge abutment. Third is a 30-minute exposure to a fire at a temperature of 1475°F. This is more demanding than a fire from a ruptured tank of gasoline. Finally, the container must not leak when immersed in water for 8 hours. This corresponds to an accident in which a truck rolls off into a creek near the road.

A series of tests was conducted by Sandia National Laboratories to verify the ability of spent fuel shipping casks to meet NRC requirements:

- A tractor-trailer rig carrying a cask was crashed into a concrete barrier at 60 mi/hr and 84 mi/hr as shown in the photo.
- A locomotive going 80 mi/hr collided with a cask on a truck, as at a crossing.
- A high-speed impact was followed by a fire.

Photograph taken a fraction of a second after collision of a tractor-trailer with a concrete wall at more than 80 mi/hr. (Courtesy of Sandia National Laboratories.)

None of the fuel casks was damaged enough to release radioactiviy, and the damage that did occur confirmed the predictions from design analysis and scale-model tests. Subsequent studies include even more unusual situations. Examples are heat from a torch of burning propane from an adjacent ruptured tank car, and crushing barrels by impact with each other within a truck that stops suddenly. Rules do not call for every cask to be tested, but a typical cask of a large number must be. Test facilities in accord with IAEA rules for radioactive transport containers are available in 12 countries around the world.

Information on transportation safety performance is available from a data base called Radioactive Materials Incident Report (RMIR), maintained by Sandia National Laboratories. Information provided by J. D. McClure and C. E. Cashwell is of interest. Most radioactive materials are carried on highways, with far fewer by air. They include industrial radiation gages, radiography sources and cameras, low-level waste, and spent fuel. As noted earlier, radioactive shipments are two percent of the hazardous materials shipments, but accidents involving radioactive materials over a 24-year period (1971-1994) amounted to only 0.8 percent. This difference can possibly be attributed to better driver training and motivation. Over that span of time, 1636 events involving radioactivity were reported, including transport accidents, handling accidents, and reported incidents. The one serious accident with A-type containers involved a truck carrying unirradiated fuel that was struck, caught fire, and burned. The most important statistics deal with transport of higher levels of radioactivity using B-type containers. Of the events, only 54 were of that type, and there was no release of radioactivity in any of them.

Preparedness for Emergencies

Despite all precautions, there is some risk in transporting radioactive materials, including wastes. To minimize risks, the safest routes can be selected. Sophisticated computer programs such as RADTRAN are available to estimate the hazard for both the accident-free mode and the accident mode. Doses to all persons that the shipment encounters are calculated, along with the accident probabilities in several categories of severity. Other programs involve the highway routes and shipping costs. Such computer programs will be used extensively in analyzing shipments of spent fuel.

An emergency response capability is available in the U.S. to cope with waste transportation accidents. The first responders to an accident on the highway would be police or firefighters. They would set up an exclusion area, attempt to save lives, and report to headquarters, starting the notification of the state emergency center. The second responder would be a state radiological protection team who would help in stabilizing the situation. In a serious emergency, governmental assistance can be quickly mobilized. The Federal Emergency Management Agency has a Federal Radiological Emergency Response Plan in concert with 11 other government agencies. They have plans to help in case of a transportation accident as well as one at a nuclear plant.* The third responder would be a commercial organization that specializes in cleaning up radioactive materials.

Federal Register, Vol. 49, No. 178, p. 35896 (1984) and Vol. 50, No. 217, p. 46542 (1985).

CHAPTER 18
Health, Safety, and Environmental Protection

The goal of safe waste disposal is to ensure that practically no radioactive material reaches human beings. Thus in managing radionuclides, account must be taken of the various ways material could be released, transferred, and absorbed. These are called "pathways to man."

Pathways

The artist's sketch shows how the dispersal of radioactivity by air, water, and land is related to the food chain of plants and animals.

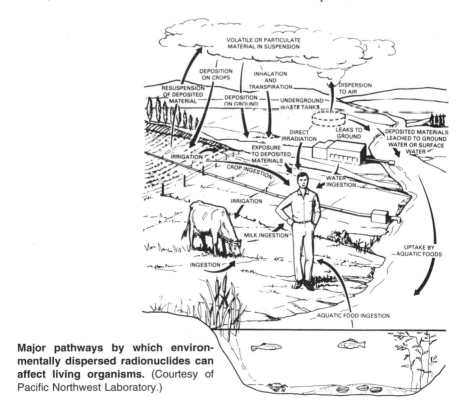

Major pathways by which environmentally dispersed radionuclides can affect living organisms. (Courtesy of Pacific Northwest Laboratory.)

The example shown is an underground source of radioactivity. If erosion by air and water removed soil and exposed the wastes, some particles could enter the air, be deposited on the ground elsewhere, and become a source of direct radiation. The material might also be deposited on plants or get into the soil in which crops are grown. Milk or meat from cattle that graze on the vegetation would then be a pathway to humans. If waste containers leaked into surface water or groundwater, nearby streams or wells could become contaminated. Water used for irrigation could have the same effect on plants as particles from the air. Drinking the water or eating aquatic foods from the stream could also be harmful.

An applicant for a license to operate a disposal facility for radioactive waste must provide evidence that pathways will not result in an excessive dose to a worker or a member of the public.

Multiple Barriers

A "systems approach" has been adopted in the design and construction of disposal facilities for radioactive wastes, whether high-level, low-level, or transuranic. That is, several obstacles are placed between the wastes and habitations. In this "defense in depth" method, the first barrier to movement of contaminants is the waste's physical and chemical form. For low-level wastes, the materials will be dried, incinerated, or compressed. They may then be mixed with a solid such as cement, asphalt, or plastic, in a noncorrosive condition.

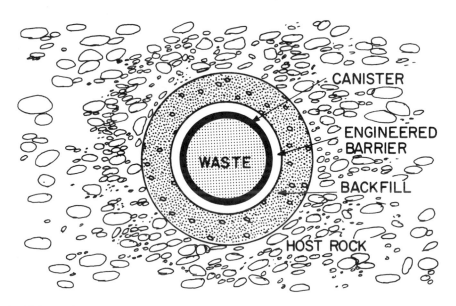

Multiple barriers to prevent waste migration. (Courtesy of Battelle Memorial Institute.)

For high-level wastes from reprocessing, the fission products may be melted with glass, as discussed in Chapter 20, or mixed with natural compounds that strongly bind the waste. For spent fuel, the uranium oxide fuel form itself is resistant to corrosion. The rods in assemblies may be consolidated to reduce the volume occupied. The second barrier is the container, which will be composed of a suitable metal, plastic, or ceramic. Corrosion-resistant metals include stainless steel, pure copper, titanium, and various alloys. More than one container may be used. Third is a fill material such as bentonite clay, which swells when it becomes moist and thus prevents the passage of water. Other physical barriers such as concrete containers or walls may also be employed. Finally, the layer that provides the most protection is the geologic medium—the soil or rock—that filters and delays particles of radioactive material from the flowing water.

Each barrier has a role in preventing migration of radioactive material. The waste form and container hold the wastes for several hundred years, during which most of the radionuclides decay to safe levels. Eventually the waste is dissolved by underground water. The surfaces within the rock, such as particles, pores, grain boundaries, and fissures, all have the effect of retarding the motion of waste material. Most of the chemical species remain within the medium. Since considerable distance is maintained between the disposal site and civilization, any wastes that remain will be highly diluted.

The Earth's Water Cycle

Since water is the principal carrier for wastes, we need to consider how the "water cycle" works. The earth's water cycle is driven by energy from the sun. As we see in the sketch, water evaporates from streams, lakes, and the ocean. It forms clouds and falls as rain or snow. Part of the water soaks into the ground, and part of it runs off, eventually reaching the rivers that carry it to the sea. Water also evaporates from the earth's surface and is released by vegetation. Thus water moves continuously from the atmosphere to the ground and back again.

A "water table" forms at some distance under the ground surface. It can be thought of as the top of the body of water that has settled into the ground. Its location varies from being near the surface close to the coast to deep below the surface in an arid inland region. Above the water table is the unsaturated zone, which contains air and some moisture; below the water table is the saturated zone. The migration of wastes is much slower in the upper region. Water tables vary somewhat with time of day and considerably with the season. A well from which water is to be drawn must penetrate the earth at least to the water table, and preferably much deeper. Pumping water from a well causes a tilt in the water surface, allowing the water to flow down the slope to match the extracted flow.

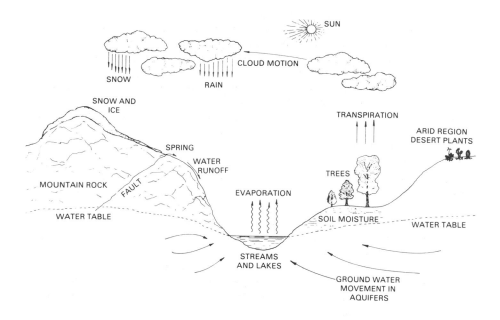

The water cycle. Precipitation falls on the earth, water flows within the ground, and evaporation completes the cycle.

Water moves within the earth by means of aquifers, which are bodies of porous rock that readily conduct water. An example is a layer of gravel. As shown in the drawing, underground water generally tends to flow toward depressions just as surface water does, but more slowly. If there is a great deal of rainfall and there are good aquifers, the water table will be near the surface. If there is little rainfall and the rock is impervious to water, the water table may be thousands of feet down. There may be more than one aquifer, separated by impervious material. Fortunately, aquifers tend to lie parallel to the earth's surface so that water flow is not directly upward.

Waste Retention in Geologic Media

The way dissolved materials are transported by underground water is very complex because of chemical and physical effects. Only by means of a computer program can the performance of a disposal facility be predicted accurately. We can illustrate how the mathematical models are applied, starting with very simple situations. Suppose that some tritium, the heaviest isotope of hydrogen, leaked from a waste container. Since tritium is practically the same chemically as ordinary hydrogen, water formed from it will move through the earth just like ordinary water. From the velocity of the water flow and the distance between the start and end of the motion, we can easily calculate the time the tritium takes to get there, as distance over velocity. That time would then be used to find out how much decay of the 12.3-year tritium has occurred during transit.

The geologic medium's ability to slow or trap radioactive material depends on the elements involved, on the chemical species, and on the electric charge of the ions in solution, as well as on the nature of the rock. A radioisotope such as cesium (30 yr) will move much more slowly than tritium. Its chemical form will cause molecules to become attached to soil particles or on pore walls in rocks as shown in the diagram. The molecules will later leave the surface and move along with the water, but the waste molecule motion has been delayed and the net speed reduced, an effect called "retardation." A retardation factor of, say, 35 means that the speed of cesium is only about 3 percent of the speed of water. The slow movement allows for decay that can reduce the radioactivity to a safe level by the time the cesium reaches public water supplies. Many of the chemical compounds formed from transuranic materials such as plutonium have very large retardation factors, and the final concentrations are low, even though the half-lives are long.

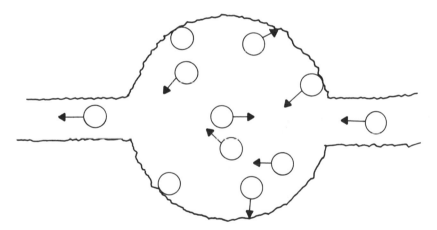

Retardation of waste chemicals in a geologic medium. The deposition of waste molecules on walls of pores on the rock delays migration.

Suppose that the waste were dissolved from the container at a constant rate over a definite period. As sketched, a "square pulse" of contamination would move along through the ground and pass a point which is farther along. The concentration of cesium, shown as the shaded area, is reduced by decay with time. Well water drawn from the aquifer at a distant point would be contaminated for a while, but would eventually clear up as the pulse moves on. A refinement of the computer model takes account of fluid-mixing effects and the nature of the geology that together cause some molecules to move more slowly while others move faster. This process, called dispersion,

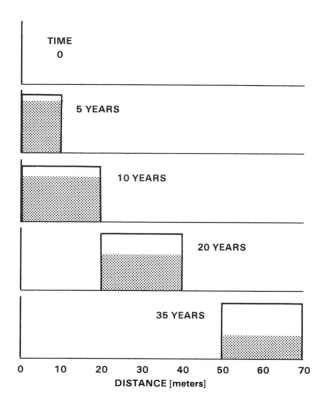

Movement of cesium-137 through the ground, showing reduction in concentration by radioactive decay. (Waste dissolving time 10 yr, waste speed 2 m/yr, half-life 30 yr.)

reduces the peak radioactivity but lengthens the time the contamination is present.

The analysis recommended by the Nuclear Regulatory Commission is more complex than we have described; it must take into account a variety of waste types and the many dissolved chemical species represented in the fission products and activation products from nuclear reactor operation. In addition, the ways the water is drawn from the ground and used by the public must be examined. Computer programs must be tested for correctness, accuracy, and applicability to the actual situation.

Safety Objectives

A frequent question about any waste disposal site is, "How safe is it?" It would be good if one could reply, "Absolutely safe, forever," as if no radioactive material ever escaped. This is not seen as a realistic answer. Instead, as discussed in Chapter 6, recommendations of standards organizations are translated into dose limits by regulatory bodies. The effects of the manmade structures and the natural barriers that inhibit the transport of radioactivity are assessed as a *system*, the performance of which is compared with the dose limits. Such a performance assessment takes into account all available information and makes conservative assumptions where data are lacking.

Nonetheless, one can estimate the time span over which calculations should be made for a high-level waste repository that accepts reprocessed fission products. Let us plot a graph of the ingestion toxicity, as measured by the number of cancers resulting from ingestion of all the waste, plotted against years after reprocessing. For reference, the practically straight dashed line is the toxicity of the original natural uranium from which the reactor fuel came. The curves cross in the vicinity of 10^4 years. We would conclude that the repository should be maintained secure for at least that long, on the grounds that the natural hazard of underground uranium ore is tolerable. The graph gives a qualitative basis for the regulatory requirement that performance of the facility for 10,000 years be assessed. Since spent fuel also contains uranium and plutonium, the time for the toxicity curve to cross the natural uranium line is still further out in time.

Another requirement is based on radiation dose. The allowed dose per year is 5 rems for a worker at a radioactive waste disposal facility. However, the ALARA principle discussed in Chapter 6 should be

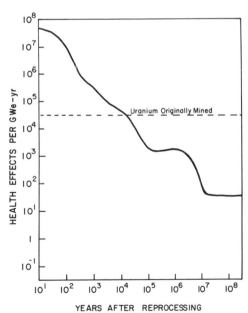

Comparison of toxicities. Toxicity from reprocessed high-level waste exceeds that of the uranium ore from which the fuel came until about 10,000 years. (Adapted from B. L. Cohen, *Am. J. Phys.* 54(1), January 1986.)

applied to keep the actual doses lower than that figure. The dose limit for people living near a site is much lower than that for occupational exposure. For any member of the public, the annual dose above background should be no more than 25 millirems for the whole body. This serves as a performance specification for the disposal site. If we use 360 millirems as the average annual dose experienced by individuals in the U.S., this limit corresponds to about 7 percent of the normal dose. Such an increase is regarded as insignificant in relation to the natural variations of dose by geographical location. For regulations on low-level waste disposal, time spans of 100 years and 500 years are used, as providing adequate time for decay; for regulations on high-level waste and spent fuel, times of 10,000 years are used because of the presence of long-lived heavy metals such as uranium and plutonium.

Natural Analogs

Models to predict waste migration can be partially checked against data collected in natural deposits of minerals. One of these is a uranium oxide deposit in Peña Blanca, Mexico. The ore is similar to the spent fuel in its corrosion characteristics. Another is a deposit of thorium in Brazil. Thorium behaves like an ion of plutonium. The rate of migration of thorium is found to be extremely small.

The most important analog is the "natural reactor" in Africa, at Oklo, Gabon. About two billion years ago, the concentration of ^{235}U in uranium was much higher than it is today. The half-life of ^{235}U is 7.04 x 10^8 years, while that of ^{238}U is 4.46 x 10^9 years. There was enough water present in the uranium ore bodies for a chain reaction to occur, with neutron bombardment, to produce fission products and plutonium. The process may have continued for hundreds or thousands of years at a low, self-limiting power level. Little migration has been noted of some radioactive species, especially ruthenium, technetium, and neodymium. Study of the amounts of different substances lost and remaining will help us to understand retention for the high-level waste repository planned for Yucca Mountain in Nevada. There are actually two types of natural reactor, one without organic materials present, the other with considerable bitumen (asphalt). The latter appears to bind wastes well, suggesting that bitumen be used in the repository.

It is an interesting exercise for the mathematically inclined reader to calculate what the atom percent ^{235}U was two billion years ago, knowing the present 0.720 percent and the half-lives of the two uranium isotopes.

Long-Term Effects

Safe disposal of radioactive wastes involves selection of materials and geology ensuring that dose limits are met over a long period of time. These requirements are part of a broader issue of meeting energy needs in the near future and in the centuries thereafter. In the present era, energy stored in the form of coal and oil is abundant, but depletion of reserves in the future is considered inevitable.

The burning of fossil fuels in electric power plants and in vehicles for transportation adds large amounts of carbon dioxide and other gases to the atmosphere. These "greenhouse gases" create a blanket that traps heat radiation and contributes to general global warming. There is disagreement as to the magnitude of the temperature rise and its consequences, especially related to the possible rise in coastal water levels. At question is the relative effect of natural phenomena versus the effect of human activities. Indications are that wide fluctuation is normal, that the climate was more nearly tropical in the era of the dinosaurs, and that several ice ages have occurred in more recent times. One popular view, however, is that when in doubt, one should take every precaution to prevent harm.

Other effects of the release of chemicals by the consumption of fossil fuels are acid rain that pollutes bodies of water and kills vegetation, and depletion of the atmosphere's ozone layer, which protects us from skin damage by the sun's ultraviolet light.

There is uncertainty as to how to cope with all of these problems. Improvements in efficiency of energy use is possible, but slow to implement, even in developed countries. There are needs for large amounts of additional energy in the developing world. It is not reasonable to expect conservation of energy if it implies reduction in its beneficial use. Greater use of solar energy and renewables is favored. However, the low energy intensity of sunlight and its variability limit its application, while the burning of firewood merely contributes to the atmospheric pollution.

In this context, nuclear power plants would appear to serve as a useful alternative to coal-fired plants. That idea is usually dismissed in discussions of the development-environment issue. Concerns are expressed about reactor safety, weapons proliferation, and the "waste problem." There is a tendency to reject nuclear without recognizing that there is a great deal of information available on performance of radioactive waste disposal systems, and that many knowledgeable scientists and engineers agree that radioactive wastes can be safely managed.

CHAPTER 19

Disposal of Low-Level Wastes

The management of low-level radioactive wastes is evolving. Early disposal practices were inadequate; stricter regulations have been developed; and designs of facilities other than shallow land burial have been proposed. The problem of finding acceptable locations for disposal sites is generally believed to be more social than technical. We shall review the history of low-level waste disposal and examine recent developments. Comments by Constance Kalbach Walker and John Mac Millan on this chapter are appreciated.

Early Practices

Only after World War II did the problem of low-level wastes arise. New radioisotopes for medicine, research, and industry became abundant through irradiation in nuclear reactors. Their residues formed wastes, and they continue to do so. The second and much larger source was nuclear power generation, which produced contaminated materials and equipment.

The Atomic Energy Commission (AEC) maintained control of all nuclear activities for nearly a quarter of a century. The map shows the location of LLW generation, storage, and disposal sites. The AEC disposed of its wastes by near-surface burial. In 1963 the first commercial disposal site was established, and by 1971 there were six sites, located at West Valley, New York; Sheffield, Illinois; Maxey Flats, Kentucky; Richland, Washington; Beatty, Nevada; and Barnwell, South Carolina. The volumes of LLW that have been placed in the six commercial sites are shown in the following table. At all of these sites, the technique basically was to dig a trench, fill it with boxes and drums of waste, replace the excavated earth, apply some compaction, and form an earthen cap above the trench. As such, they were a cut above sanitary landfills, but below modern standards for disposal facility design.

Shallow Land Burial

The disposal of LLW in a trench slightly below the surface is called shallow land burial, in contrast to deeper mined-cavity disposal as practiced for high-level wastes. The diagram shows the conventional

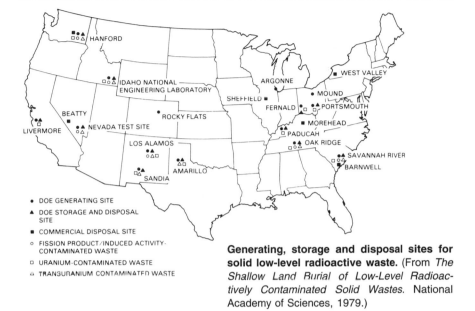

- ● DOE GENERATING SITE
- ▲ DOE STORAGE AND DISPOSAL SITE
- ■ COMMERCIAL DISPOSAL SITE
- ○ FISSION PRODUCT/INDUCED ACTIVITY-CONTAMINATED WASTE
- □ URANIUM-CONTAMINATED WASTE
- △ TRANSURANIUM CONTAMINATED WASTE

Generating, storage and disposal sites for solid low-level radioactive waste. (From *The Shallow Land Burial of Low-Level Radioactively Contaminated Solid Wastes.* National Academy of Sciences, 1979.)

Volumes of Low-Level Wastes at Commercial Burial Sites
(Adapted from Oak Ridge National Laboratory
report DOE/RW-0006, Rev. 9, March 1994)

Site	Volume, in millions of cubic feet
Barnwell, South Carolina	23.33
Beatty, Nevada	4.34
Richland, Washington	11.95
Maxey Flats, Kentucky	4.78
Sheffield, Illinois	3.12
West Valley, New York	2.72
Total	50.24

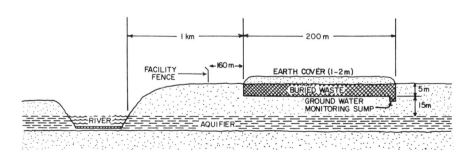

Shallow land burial of low-level wastes. (Courtesy of R. F. Weston, Inc.)

arrangement, with the wastes placed a suitable distance above an aquifer and the site fenced to prevent entrance. There has been mixed success with this disposal technique. Three of the commercial sites (West Valley, Sheffield, and Maxey Flats) developed leaks and were closed, while the remaining three have operated satisfactorily.

Three principal types of failure have been noted.* The first type is simple erosion by surface water, which exposes waste containers to the elements. The second type occurs when the wastes are loosely packed and will compress under the weight of dirt. Then the cap subsides, pockets of water appear, and water percolates into the waste, eventually leaching out the radioactive material. The third type is the "bathtub effect," in which a leaking cap lets water into an excavated cavity with nearly impermeable walls. The wastes are immersed in water for long periods, the containers corrode, and the wastes dissolve. The "bathtub" fills and overflows, carrying contaminated water to the environment.

Accounts differ widely as to the amounts of radioactivity released and the resultant hazard to the public. It is a fact, for example, that tritium and strontium-90 have been measured at Maxey Flats. One estimate is that the resulting dose was less than 1 percent of that from natural background. Statements by others suggest that the hazard was much greater.

The design of such earlier facilities was clearly inadequate. The problems were due partly to insufficient investigation of geologic features before sites were selected. Contributing factors were the loose packing of wastes, the presence of liquids in the waste as received for disposal, and the poor design of caps to exclude water. Operating companies claim that they followed the federal regulations existing at the time. Hindsight reveals the primitive nature of regulations on site selection, waste packaging, and facility design.

The remaining sites (Richland, Beatty, and Barnwell) had problems of a different type. Poor packaging of wastes by shippers resulted in contamination of the area on arrival at the sites. At one location, contaminated equipment was stolen. The three states were concerned with the injustice of having to receive the bulk of the wastes from the entire nation. As a result, the governors of the states threatened to close the facilities, leaving no place for LLW to go. This prompted Congress to pass in 1980 the Low-Level Radioactive Waste Policy Act, which placed responsibility for waste disposal on the states producing the waste. The Act also recommended that groups of states establish regional facilities. Several interstate compacts have been arranged, as discussed in Chapter 22.

---

*See "Proceedings of the Symposium on Low-Level Waste Disposal," Vols. 1-3, NUREG/CP-0028, 1982-3.

Site Selection

Experience at the closed disposal sites led the Nuclear Regulatory Commission to develop a new set of siting rules. The result after several years of work was the regulation entitled "Licensing Requirements for Land Disposal of Radioactive Waste," Title 10 of the Code of Federal Regulations, Part 61, commonly referred to as 10 CFR 61. The regulation requires low-level wastes to be classified by the generator into one of three categories—A, B, or C, taking account of half-lives and concentrations of the nuclides in the wastes. Generally, Class A wastes are of lower activity than Class B or Class C. The latter require special attention, as will be discussed in Chapter 22. Regulation 10 CFR 61 stresses the need for careful assessment of geology, hydrology, and other features. Some key requirements are (a) the site geology should be simple enough to admit mathematical modeling and computation, (b) location of the site should be far from housing or commercial development that would affect underground water flow, (c) there should be no significant underground resources whose extraction would affect site performance, (d) the sites should be well-drained to reduce infiltration of water, (e) there should be little chance for water flow across the site that would erode the surface, (f) the water table at the site should be deep enough to prevent immersion of wastes, (g) there should be no springs in the site that could bring contamination to the surface, (h) the area should not be subject to volcanic action, earthquakes, landslides, or excessive weathering, and (i) nearby activities should not affect performance or monitoring.

The process by which a site is selected consists of three parts. The first is a general survey of the region or state, examining existing records on the geology, hydrology, meteorology, seismology, population distribution, land use, environmental features, and cultural aspects. The second is an interaction between the siting agency or company with communities that may wish to serve as host to the facility or at least will find it acceptable with appropriate compensation. The third is a series of measurements, including a limited number of test borings, leading to a site characterization, which is a catalog of features that relate to the siting regulations.

The site characterization study following NRC guidelines must be carried out before a license to operate can be issued. Such an investigation must include data taken over a full year. Once a license is issued, the disposal site can be opened to receive wastes, and will operate for about 30 years, during which regulatory surveillance is provided by the NRC or the cognizant state agency. At the end of the operating period, the site may be closed, meaning that no additional wastes are received. However, institutional control must continue for at least an additional 100 years. This involves measurements of waste migration and remedial action as needed. After institutional control is released, the site may be used safely for any purpose. The design is expected to protect the public for at least 500 years after closure. The performance specification of no more than 25 mrems per year to any member of the public must be continued.

Greater Confinement Disposal Alternatives

The NRC has suggested several techniques by which the 10 CFR 61 requirements might be achieved, starting with conventional shallow land burial with sloped trench walls. Use of a narrow trench with vertical walls cuts down on the radiation exposure to operators as they place the wastes. Containers can be stacked in a regular pattern to reduce the amount of potential void space. Cubes or hexagons leave essentially zero voids. Square 55-gallon drums are now available commercially. Layered waste disposal puts material of high activity in the bottom of the trench to protect the inadvertent intruder. The NRC recommends several moisture barriers as part of the cap system. It also suggests a design for engineered barriers against intrusion.

The three disposal facilities at Richland, Beatty, and Barnwell did not experience difficulty with waste migration and were regarded as successful even though they are examples of shallow land burial. Those in the West are in arid regions. Barnwell is in a humid region with heavy rainfall but is located where much of the soil is clay, which prevents water intrusion and filters contaminants effectively.

Despite these favorable conditions, publicity about the now-closed LLW disposal sites has given shallow land burial a bad reputation. The public is concerned about reactors, radioactivity, and radiation in general, and shallow land burial in particular. Consequently, some states and compacts have prohibited its use or have required an "alternative waste disposal technology" that involves "greater confinement disposal." Several design concepts that incorporate "engineered barriers" have been proposed. Generally, these provide additional protection against waste migration.

The Nuclear Regulatory Commission finds that a facility designed to meet its performance specifications is satisfactory without any added features. The public often seems to have one or more of these different views: (a) the limiting dose should be nearer zero or even should be actually zero, (b) some unexpected event might change the system from the one analyzed, (c) the knowledge of underground water flow is inadequate, and (d) there may be human error in the analysis, design, construction, and operation of the facility. It is hard to convince people otherwise, and in some states and interstate compact regions, legislation on additional protection has been passed to make a waste disposal facility acceptable.

We can note some advantages and disadvantages of several of the proposed concepts. Referred to are the accompanying sketches, kindly provided by EG&G Idaho, Inc., of the Idaho National Engineering Laboratory, as part of DOE's National Low-Level Waste Management Program.

Intermediate-depth disposal is similar to shallow land disposal except for the greater trench depth and except for the thickness of the cover, which may be more than 30 feet. The cover minimizes water infiltration, provides ample shielding against radiation, and discourages future intrusion.

Intermediate-depth disposal. Similar to shallow land burial except for deeper trench and thicker cover. (Courtesy of Idaho National Engineering Laboratory; artist David Combs.)

Mined-cavity disposal consists of a vertical shaft going deep in the ground, with radiating corridors at the bottom. This is quite similar to the planned disposal system for spent fuel and high level wastes from reprocessing. This method is not likely to be used in the U.S. because of the greater expense. We note, however, that Finland uses cavities cut in bedrock for LLW disposal. Deep disposal can be used for wastes not suitable for near-surface disposal.

The *belowground vault* provides a concrete barrier to infiltration of water and to the escape of radioactive materials. Reinforced concrete forms the walls, floor, and roof, with waste containers stacked closely together and the spaces between filled with sand or gravel. The vault enhances stability under the pressures of overburden and protects the waste against effects of earthquakes. It sits on a porous pad to allow any water that enters to drain away. The structure also provides gamma-ray shielding for those at the surface and minimizes future intrusion. Concrete with an average life of around 500 years is required, using special mixes and methods of preparation. The earthen cap has several layers: (a) vegetation that is hardy but has shallow roots, (b) topsoil, (c) rocks to prevent root entrance, (d) sand and gravel to carry water sideways, (e) clay to provide a barrier nearly impervious to downward water flow, and (f) sand and gravel to divert water. The cap also provides shielding and deters intrusion.

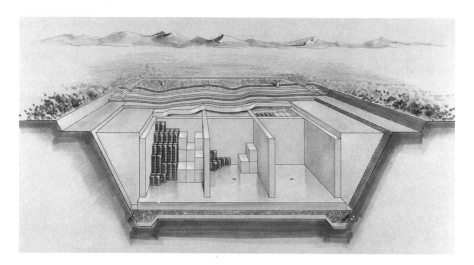

Belowground vault. Concrete structure with earthen cover. (Courtesy of Idaho National Engineering Laboratory; artist David Combs.)

Modular concrete canister disposal consists of placing individual waste containers within concrete cylinders called canisters, with grout filling the spaces between containers. The canisters are then buried below grade in a shallow land site, with sand between them for stability and with the usual earthen cover.

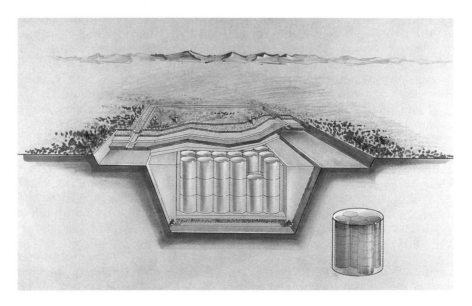

Modular concrete canister disposal. Waste placed in canisters which are placed in trenches. (Courtesy of Idaho National Engineering Laboratory; artist David Combs.)

Earth-mounded concrete bunkers, used by the French at a recently filled facility, have the favorable features of the belowground vault and the modular concrete canister. Well below the natural grade a concrete bunker is constructed to hold the more radioactive Class B and C wastes. The mound above, called a "tumulus," is formed by a stack of Class A concrete containers, of lower activity. A rounded earthen cover allows runoff, prevents infiltration, and provides radiation shielding.

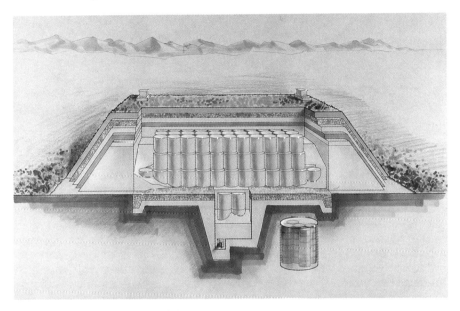

Earth-mounded concrete bunker. Class B and C wastes in bunker with Class A wastes in mound above. (Courtesy of Idaho National Engineering Laboratory; artist David Combs.)

Shaft disposal uses concrete for walls and a cap. A series of individual holes are drilled in the ground and concrete cylinders are installed, into which waste containers are placed. The method is seen to be a variant on either the modular concrete canister or the belowground vault.

The *aboveground vault* is a simple reinforced concrete box into which containers of waste are placed, with sand between containers to ensure stability. A concrete slab is poured in place above each cell after it is filled with waste containers. If necessary, the waste containers could be retrieved. Thus the concept is more nearly "storage" than "disposal." These words have significant overtones in terms of public acceptance. Storage is favored by those concerned about the presumed hazard of burial of waste with future concealed problems; disposal is favored by those who believe that later generations should not be burdened with long-term custodial responsibility.

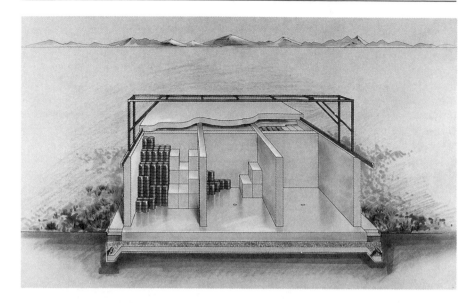

Aboveground vault. Concrete building without earthen cover. (Courtesy of Idaho National Engineering Laboratory; artist David Combs.)

As no earthen cover is provided, the concrete roof is exposed to effects of weather, including erosion, freeze-thaw cycles, and acid rain. Of course, deterioration of the structure could be noted, and repairs made, but surveillance would need to be extended, perhaps indefinitely. Analyses show that if the aboveground vault is unattended, the concrete would deteriorate and the wastes eventually be dissolved and transferred by surface water, giving an unacceptable radiation dose. The concept's features are not particularly compatible with the requirements on protection of an inadvertent intruder, who encounters the facility beyond the end of the 100-year institutional control period. For these various reasons, the NRC does not favor this method of disposal, as noted in Volume 3 of the series of NRC reports on alternatives (see References, Appendix B).

A concept that combines features of others and goes considerably further in terms of protection is the *Integrated Vault Technology*, designed by Chem-Nuclear Systems, Inc., for use at several U.S. disposal sites. The design borrows from the technology used by the French at Centre de l'Aube. As sketched, earth is mounded above each module, with a sloping multilayer cap to help runoff. A plastic layer is added to reduce infiltration of water. The facility uses reinforced concrete vaults, into which concrete canisters called "overpacks" are placed, each holding one or more low-level waste containers. Overpacks are of two types. One is cylindrical, into which a number of 55-gallon drums may be fit, with spaces between filled with grout. Alternatively, a single high-integrity container can be inserted. The other is rectangular to accommodate square liners or a set of

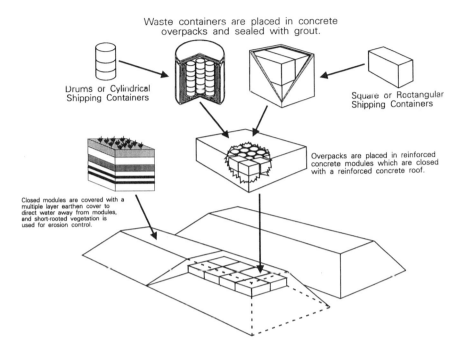

Waste containers are placed in concrete overpacks and sealed with grout.

Drums or Cylindrical Shipping Containers

Square or Rectangular Shipping Containers

Overpacks are placed in reinforced concrete modules which are closed with a reinforced concrete roof.

Closed modules are covered with a multiple layer earthen cover to direct water away from modules, and short-rooted vegetation is used for erosion control.

Integrated Vault Technology. (Courtesy of Chem-Nuclear Systems, Inc.)

boxes. The use of overpacks in addition to the module vault provides structural stability and an additional layer of protection. All classes of waste are put in overpacks. The higher activity B and C wastes are placed in the bottom of the vault, with A wastes above and beside them. This arrangement provides shielding and protection of future inadvertent intruders. In contrast with other concepts, spaces between overpacks are not filled with sand or gravel, but are left open to allow free drainage of any infiltration. The vaults sit on porous pads for drainage, with the bottom of the disposal facility well above the water table. Each vault has a drain to a monitoring system, located in a tunnel between adjacent rows of vaults. The tunnel allows access for inspection, sampling for possible water drainage from the vaults, and measurement of any radioactivity. Such a facility is obviously much more expensive than simple shallow land burial or other variants, and it is expected to amply meet the stringent requirements of both the NRC and Agreement States, even for a facility located in the humid region.

We can visualize how a facility using the Integrated Vault Technology provides protection of the public. The sloped multilayer cap allows runoff of a large fraction of rainfall. Most of what infiltrates is drawn sideways by the sand-gravel layer, and the combination of plastic and clay beneath prevents infiltration. The concrete roof and its contents thus remain dry for many years. During the 100-year institutional control period, which can be longer if desired, remedial action can be taken if significant amounts of water enter any vault or if radioactivity is detected by the monitoring system. The plastic in the cap is assumed to deteriorate after a number of years, and some water will enter, contacting the vaults. Eventually cracks will appear in the concrete roof and walls. Moisture can then contact the outside of the concrete overpacks, but because of the open spaces between them, they are not exposed to standing water. Over a period of many hundreds of years, the overpacks are assumed to deteriorate, gradually allowing water to contact the waste containers. The concrete roof and overpacks still effectively warn any inadvertent intruder and provide a physical barrier to exploration. By this time, much of the radioactivity has disappeared by decay. Finally, water diffuses into the waste form, slowly dissolves some of the chemicals of the waste, and diffuses out into the vault, to be carried away in the ground beneath the facility. As in any of the disposal concepts, each chemical migrates with a speed slower than that of the underground water by an amount called the retardation factor, values of which vary from one to thousands. Knowledge of the water flow patterns and velocity allow prediction of the amount of decay in transit through the buffer zone, which is a minimum distance (e.g., 1000 to 2000 feet) between the facility and the nearest public water well. The facility design and the geologic environment act as a *system* that stores and isolates the wastes and delays the release of radioactivity. Concentrations of activity are not zero but are expected to be well below the regulatory limit during operation and for all of time after the facility is closed.

The cost of disposal of low-level waste continues to increase for at least two reasons. First, legislators and regulators have responded to public demand by requiring more secure and complex disposal facilities. Second, in the interest of economy, generators have actively sought to reduce volumes of waste to be shipped and disposed of. Savings are not nearly as large as expected because much of the expense of disposal comes from the facility's initial capital cost, which may be over $100 million. A significant part of the initial cost comes from locating and characterizing sites under existing political and regulatory requirements. A number of fixed facility operating costs are independent of waste volume; consequently, any reduction in volume by generators increases the charge per cubic foot. DOE has estimated that for a facility with reference input of 250,000 cubic feet per year a reduction in volume of 75 percent would result in a 300 percent increase in cost per cubic foot. For disposal in facilities that serve a small number of customers, such costs might be ten times higher than for a large facility. A portion of the income from waste disposal goes to the community or state that serves as host; another part goes into a fund for remediation if needed; and finally money is collected for the design and development of the next facility to serve the state or region. Generators pass the costs on to the customers who use nuclear electricity and to the beneficiaries of medical and research applications. The public ultimately pays the bill, of course.

Special Classes of Wastes

Most low-level wastes come from nuclear reactor operations and radioisotope applications, regulated by the NRC. One type that is the responsibility of states is NARM (naturally-occurring and accelerator-produced radioactive material). The EPA rule 40 CFR 193 will address such substances as well. Part are accelerator targets and other equipment. Others are soils containing natural uranium or radium. The latter element was produced naturally in large quantities in earlier times, as described in the *Scientific American* article "The First Nuclear Industry," listed in Appendix A. Radium contamination exists at former plant sites. Such wastes may be disposed of as LLW or sent to a desert disposal facility in Utah operated by the company Envirocare.

"Mixed wastes" are those containing both hazardous chemicals and radioactive substances. Hazardous wastes are defined as materials that are toxic, corrosive, inflammable, or explosive. They contain specific elements such as lead and mercury, pesticides such as DDT, and cancer-producing compounds such as PCBs and dioxin.

The disposal of hazardous wastes is regulated by the Environmental Protection Agency under the Resource Conservation and Recovery Act (RCRA) while radioactive wastes are controlled by the Nuclear Regulatory Commission under the Atomic Energy Act. Thus it has been necessary to establish consistent dual rules by agreement between agencies. The EPA and NRC provide a formal procedure by which one can decide whether a certain material is mixed waste. They also give written suggestions for the design of a disposal facility that meets both agencies' requirements. Included are double liners and leachate recovery equipment for EPA and waste isolation and intruder barriers for NRC. It is estimated that only a few percent of commercial and institutional low-level wastes are in the category of mixed wastes. Their disposal, however, is estimated to be very expensive, as high as $15,000 per cubic foot, which is about 100 times that for ordinary low-level waste. Since the total volume of mixed wastes is relatively small, they do not pose a major problem, but their management has to be addressed.

Generators of mixed waste have a sequence of possible options: (a) to select alternative materials and processes that do not involve hazardous organics or metals, leaving only radioactive materials to deal with, (b) by operational control to reduce the radioactive content below NRC concerns, leaving only the hazardous component to deal with, (c) to store wastes for decay if half-lives are short enough, (d) to ship wastes to a processing facility which can treat them to eliminate the hazardous chemical property, leaving only the radioactive component, and (e) to ship the wastes to a jointly regulated facility.

States and compacts in turn have options. One is to build and operate a separate special disposal unit for mixed wastes, one that meets all regulators' requirements. Another is to seek relief by requesting the Department of Energy to accept mixed wastes for treatment and disposal as a small fraction of DOE's accumulation. DOE is completing a National Mixed Waste Profile, an inventory of amounts of mixed wastes in storage and being produced, and an identification of facilities capable of processing them. Such data are useful for defense waste disposal, to be discussed in the next chapter.

Another special waste is given the name "below regulatory concern (BRC)," as being materials so slightly contaminated by radioactivity that it would be safe to dispose of them as ordinary waste, e.g., in sanitary landfills. No agreement has been reached among the NRC, industry, and the environmental community as to an acceptable method of handling such wastes.

"Greater-than-Class-C" (GTCC) wastes are in the category of low-level wastes but differ in two ways. They either have a higher activity than the upper limit of the NRC's Class C or contain larger amounts of transuranic materials than allowed. Examples are activated metal wastes from reactor decommissioning and TRU wastes from nuclear fuel testing or uranium-plutonium mixed oxide fuel fabrication. Such materials cannot be given near-surface disposal. The Department of Energy is committed to accept GTCC wastes for storage and disposal along with high-level wastes. The volume of such wastes generated through the year 2020 is estimated to be relatively small.

CHAPTER 20
Disposal of Defense Wastes

Nuclear technology seems relatively new, but the year 1992 was the fiftieth anniversary of the startup of the first nuclear reactor. For over a half-century, radioactive defense wastes have been generated, first in the World War II atom bomb project, and ever since in producing materials for nuclear weapons and in reprocessing fuel from nuclear-powered naval vessels. Given the national policy to maintain a strong nuclear defense as part of the Cold War, the production of wastes was inevitable.

The defense products have always had the highest priority, and wastes have been handled in ways that seemed appropriate at the time. Thus decisions were made to store waste in temporary single-walled underground tanks, as at Hanford, or to bury them in ways regarded as inadequate by modern environmental standards, as at Idaho Falls. Waste management plans must include remedial action as well as current needs.

We can appreciate the complexity of the defense waste problem by the mere listing of its many dimensions. First are the sources: from treatment of fuel from several types of reactor—those for production of plutonium and tritium, naval reactors, and test reactors—and from weapons production and research programs. Second are the geographic locations of the laboratories that produce waste; the major ones are Hanford, Savannah River, Idaho Falls, Oak Ridge, and Los Alamos. Third are the types of waste—high-level (HLW), low-level (LLW), and transuranic (TRU). Fourth are the forms in which the wastes are currently stored. Fifth are the various forms into which the waste can be put—capsules, cement-waste mixtures, and glass-waste mixtures. Sixth is the status—buried, stored, or freshly produced. Seventh are the possible ways to dispose of the waste—stored for future decision, immobilized in its present location, shipped to a special repository, or prepared for disposal in the planned commercial repository. The stated purpose of the defense waste program of the Department of Energy is to provide safe and economical management of waste to protect the environment and ensure safety and health of workers and the public. In the following sections we describe some of the key technical elements of the process and outline the strategy that is under way.

Waste Form

The wastes from reprocessing to remove plutonium and uranium have been generated since 1944. They are composed of a great variety of chemicals, including the fission products, metals such as iron, manganese, and aluminum, sodium salts, organic materials, and special radionuclides strontium-90 and cesium-137. They are in the form of sludges (of consistency from thin paste to peanut butter), solid salt cake (from jelly to concrete), and liquid, all traditionally stored in underground tanks, especially at Hanford and Savannah River.

As noted in Chapter 11, there are enormous volumes waiting to be processed into a safer form. The diagram shows such a tank with its

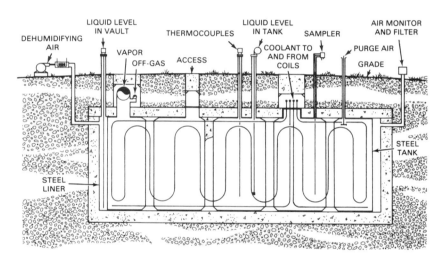

Defense waste storage tank. (From *The Safety of Nuclear Reactors and Related Facilities,* WASH-1250, U.S. Atomic Energy Commission, July 1973.)

various openings and measuring devices. The 149 early tanks at Hanford were single-walled, and in the 1960s about 67 of them developed leaks. As much as a million gallons of contaminated liquid escaped. The nature of the soil in southeastern Washington State has provided protection to the public. A National Academy of Sciences report (see References, Appendix B) notes the sorptive character of the layers of gravel, sand, silt, and clay, and states that the radioactivity has been essentially immobilized. Under the report's worst-case scenario, radiation doses at the Columbia River are predicted to be very small. Later 28 double-walled tanks were built. On detecting a leak in the inner wall, there is ample time to transfer the contents to an intact tank. An extensive ground-monitoring system tracks movement of contamination from past leaks. This includes sampling of the contents of the unsaturated zone as well as water measurements below the water table.

Considerable attention has been given to the possibility of an explosion in the Hanford waste tanks. To extract ^{137}Cs and ^{90}Sr, the chemical ferrocyanide was used. Its residue can react with other chemicals to generate heat. Also, hydrogen is released in certain reactions and can build up to hazardous concentrations. To forestall such an event, a large 20,000-lb test pump was developed. It stirs the mixture in a tank, allowing hydrogen to come out continuously rather than accumulating. Tests on the pump in Tank 101-SY have been regarded as very successful. A permanent pump is planned, with the test pump moving to other tanks.

Tank storage has to be regarded as temporary, and it is necessary to process the wastes in preparation for ultimate disposal. We can describe the steps required at Hanford and Savannah River. First is characterization, which is the sampling and analysis of the materials to determine their physical, chemical and radiological nature. In the early days, records on radioactivity were better than those on chemical content. With 177 tanks to deal with as at Hanford, extracting and analyzing even a few samples per tank is a major task. The second step is retrieval. Pumping will work well for liquid, but special excavating tools will be required for hardened material. Robot arms are being considered for this application. Third is pretreatment, in which high-level waste is separated from low-level waste and from nonradioactive materials. Water containing small concentrations of radioactivity is mixed with cement and fly ash (as from burning of coal in power plants), to form what is called "grout" at Hanford and "saltstone" at Savannah River. These are poured into large concrete vaults hundreds of feet long for permanent disposal as low-level waste.

The high-level component of the waste is to be immobilized by mixing it with material that forms a solid resistant to attack by water or the solutions that may be present underground. The substance should contain an adequate fraction of waste as impurity and still remain strong and uniform in composition. It should not be damaged by heat or radiation from fission product decay.

Glass has been studied extensively and is the material selected for most defense wastes. Glass has several favorable features: it mixes well with wastes of various compositions, casts easily into proper form, conducts heat well, and resists attack at low temperatures. Some other properties are its noncrystalline (amorphous) form, in contrast with substances such as ice, salt, and sugar. Use of the word "crystal" to describe a form of glass is a misnomer. It is in a vitreous condition as a supercooled liquid that is subject to fracture. Under excessive heat, stress, or radiation, glass could devitrify and break into pieces. The smaller the fragments, the more easily the glass is leached (dissolved) by water. The principal chemical in glass is silicon, an ingredient of sand. Typical commercial glass as used in bottles or windowpanes has this composition:

Compound	Percent
Silica (silicon dioxide SiO_2)	71.5
Soda (sodium oxide Na_2O)	14.0
Lime (calcium oxide CaO)	13.0
Alumina (aluminum oxide Al_2O_3)	1.5

The glass-waste mixture used by the French has a composition that is more like Pyrex®, the heat-resistant glassware used for cooking made by Corning. It is a borosilicate type, in which boron oxide replaces lime. When fission products are included, the glass is black rather than transparent and has this composition:

Compound	Percent
Silica	42.5
Soda	14.0
Boron oxide	17.5
Alumina	8.5
Fission product oxides	13.0
Other	4.5

The vitrification process that has been developed for HLW by the Pacific Northwest Laboratory starts with liquid waste. It is mixed with glass-forming chemicals, and the combination is introduced into a melter, as shown in the drawing. The mixture dries and the resultant solid melts. The equipment is operated remotely, behind a concrete wall 4 feet thick. At the temperature of 1150°C, organics are destroyed and inorganics dissolve in the glass. Heat is generated in the molten glass by a large electric current. A turntable allows a canister to be filled with glass, then moved around to cool, bringing up an empty canister. Gas from the process is treated to remove water, heat, and radioactivity such as carbon-14, ruthenium, and iodine-129. The canisters are welded shut, decontaminated, and stored in racks or air-cooled vaults, awaiting disposal in a repository.

At Hanford, one or more large-scale vitrification plants based on the above research are planned. These will solidify the high-level waste now in the many underground tanks. The exact timetable for completion of the vitrification facilities is not yet fixed.

At Savannah River, high-level wastes consisting of 34 million gallons and 538 megacuries of activity are to be treated in the Defense

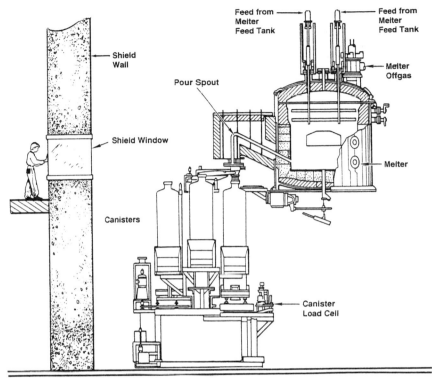

Pour Turntable

Melter and turntable designed for the Hanford Waste Vitrification Plant. The molten waste-glass mixture solidifies within the metal canister. (Courtesy of the Pacific Northwest Laboratory.)

Waste Processing Facility (DWPF) and its supporting facilities, under construction at a cost estimated to be around $4 billion. The schedule for facility completion has slipped and costs have increased due to management problems, changes in plans, and compliance with new regulations. The vitrified HLW is to be stored in stainless steel cylinders 2 feet in diameter by 10 feet in height, for eventual disposal at a federal repository.

Much research has been done on glass as a waste form. The influences of mechanical stress, radiation, and corrosion have been examined. Of special interest is the effect of water, saltwater brine, and bitterns, which are brackish residues after salt has crystallized out of saltwater. The amount of leaching depends on the element. For example, the amount removed per day in micrograms per square centimeter is 1 for cesium, 0.1 for strontium, 0.01 for cerium, and 0.001 for ruthenium.

Other materials can serve as a backup. Supercalcine consists of natural minerals that resist attack. A form called "SYNROC" (for synthetic rock) has three crystalline structures that are especially stable against corrosion.

In Situ Vitrification

A novel spinoff of the above solidification process that may be useful in stabilizing certain types of radioactive waste is "in situ" (in place) vitrification (ISV). Developed at the Pacific Northwest Laboratory, ISV converts contaminated soil or other earthen materials into a glasslike solid by electric joule heating.

Four electrodes in a square array are placed slightly into the soil to be treated. A starter path of graphite plus glass is laid down between the electrodes. Four thousand volts of electric power is applied to the starter path; this heats up and melts the adjacent soil, which, upon becoming molten, becomes electrically conductive. The molten soil then becomes the conductor for the heating process while up to 3.5 megawatts of power (roughly the amount consumed by a good-sized hotel) are applied and the melt is heated into the 1600 to 2000°C range. The molten mass grows downward and outward past the electrodes until the desired volume is encompassed. The electrodes are progressively lowered while the melt grows downward.

A glassy and microcrystalline solid similar to volcanic obsidian results upon several months of cooling. Individual melts as large as 1400 tons and 22 feet deep have been made in a single setting of electrodes. A significant volume reduction of 25 to 45 percent occurs with most soils because of the elimination of void volume. The vitrified product is totally free of organic contamination because organics are thermally decomposed by exposure to the melt temperature. Heavy metals like plutonium, uranium, and lead become permanently immobilized by chemical incorporation into the vitrified material. Leach rates are significantly smaller than those for vitrified HLW, primarily due to the higher silica content associated with contaminated soil applications. It is estimated that the vitrified product will retain its burden for geologic time spans.

ISV has been developed also for soils contaminated with hazardous chemicals. The process has proven effective for a broad range of organic and heavy metal contaminants. It is particularly competitive for more difficult applications involving mixtures of these waste types, or where application in place is important. Sublicensed to Geosafe Corporation, the technology is being marketed worldwide, and commercial large-scale applications began in mid-1993. Current projects deal with mercury, pesticides, dioxin, PCP, PCBs, plutonium, uranium, beryllium, and other organic and heavy metal wastes.

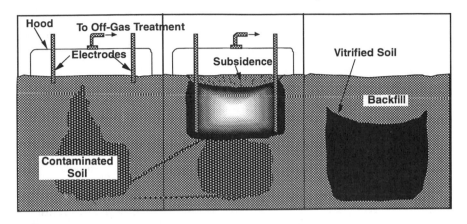

In situ vitrification. (Courtesy of the Pacific Northwest Laboratory.)

Disposal Strategy

Many types of defense wastes at many sites need to be disposed of in accord with requirements of the Environmental Protection Agency. Priorities must be established in order to accomplish the task at reasonable cost. For a number of years the Department of Energy has been implementing a long-range management plan. A comprehensive discussion of the thinking as of 1987 appears in a report, "Program Implementation Plan," DOE DP-0042. In 1989, the program was broadened in scope to include environmental restoration, and subsequently several Five-Year Plans have been issued. An important feature is prioritization, in which problem sites are attacked in order of importance and urgency. The high-level and transuranic wastes at three laboratories will be disposed of in sequential order. The Savannah River Plant (SRP) has the greatest fraction (59 percent) of the DOE radioactivity. With its wet climate and small depth to the water table, the SRP has the greatest potential for release of buried wastes. Those will be handled first. Information so gained will be useful for the Hanford Site, and in turn for the Idaho National Engineering Laboratory (INEL).

Transuranic waste poses a special problem because of its peculiar history. Prior to 1970 it was buried as LLW. In 1970 TRU was defined as wastes containing 10 nanocuries per gram of alpha-emitters (redefined in 1982 as 100 nCi/g). Since 1970 it has been kept separate from LLW and stored in underground bins. Three categories require different treatments: contact-handled (CH-TRU) with surface dose rate less than 200 mrems/hr; remote-handled (RH-TRU); and special (SP-TRU), which includes large contaminated pieces of equipment. About 99 percent is the CH type.

The general plan is to send TRU to the Waste Isolation Pilot Plant, described below, once it has been checked for its properties. TRU is transported in the special container called TRUPACT II, as described in Chapter 17.

The low-level waste from the three laboratories and other DOE sites will be buried in some form of greater-confinement disposal. Choices include those for commercial low-level waste (Chapter 19) along with lined trenches, caisson, or augured holes.

Waste Isolation Pilot Plant

In 1957 a committee of the National Academy of Sciences and the National Research Council reported the results of their study entitled *The Disposal of Radioactive Waste on Land*. The group said that it was "convinced that radioactive waste could be disposed of safely in a variety of ways and at a large number of sites in the United States." They indicated that waste storage in tanks was safe and economical for the present but that their first choice was solidification followed by underground burial. Their second choice of method of permanent disposal was use of silicate bricks in surface repositories or dry mines. Their main recommendation was that "disposal in salt is the most promising method for the near future." The salt recommended was natural sodium chloride in rock form, deposited long ago when oceans that covered the U.S. dried up. It appears in two arrangements: "bedded" or layered, and "domed," as a hill.

Salt has several virtues as a geologic medium. It is abundant, distant from earthquake zones, and inexpensive to excavate. It is almost impermeable since it is plastic—cracks and crevices are sealed by pressure. The committee believed that wastes deposited in salt would be free of contact with water in the future because the very existence of extensive salt deposits indicates past freedom from water.

The favorable opinion of salt led the Atomic Energy Commission to undertake a 10-year investigation named Project Salt Vault. Tests were made in a mine formerly used by the Carey Salt Company near Lyons, Kansas. The mine was cleaned up, repaired, and equipped with a shaft hoist, a ventilating system, and an electric power system. An underground transporter was brought down to the tunnels. Studies included the effect of heat supplied electrically, and the effects of both heat and radiation using spent fuel. Corrosion measurements showed ordinary steel to be much better than stainless steel in a salt environment. The tests were successful and it was concluded that salt was indeed a good medium. The AEC plan to put a repository in Kansas was canceled in 1970, however, when previous drilling for oil and gas was discovered and that large amounts of water had been pumped in to remove salt.

Despite this experience, salt continues to be a promising disposal medium. The Waste Isolation Pilot Plant (WIPP) is a repository under construction in ancient salt beds in southeastern New Mexico, 25 miles east of Carlsbad Caverns. Its purpose is to test the feasibility of safe disposal of TRU, of which there are some 185,000 cubic meters stored at DOE sites around the country. The whole WIPP site comprises more than 10,000 acres, most of which is buffer. The facility will have surface waste-handling facilities on an area of 30 acres. The site is on a bedded salt layer 2000 feet thick, located 850 feet below the surface. Wastes would be placed some 2150 feet down.

Work since 1982 includes the drilling of four shafts and excavation of major tunnels and seven of the planned 56 rooms. Environmental data have been collected and geological measurements made related to performance assessment. DOE had planned to start in 1991 to emplace drums of TRU for test purposes, but the project was delayed by legal action related to the withdrawal of New Mexico land for the project. Congressional action in 1992 released the land but added numerous requirements, including the need for certification by the Environmental Protection Agency of the future safety of the facility. Testing may take until the year 2000. Meanwhile, containers of TRU are deteriorating. It may be necessary to repackage the waste or convert it to another waste form, at great expense and some potential radiation hazard to workers. One of the requirements of TRU disposal at WIPP is accurate knowledge of the contents of waste containers, including materials, radioactive content, and radiation levels. Some difficulty is being experienced in establishing representative sample drums for test purposes.

Of special technical interest is the movement of salt brine, which tends to migrate under the influence of pressure and temperature. Studies will also be made of the release of gases as the result of interaction of salt with containers and waste. Continued assessment will be made of the expected long-term (10,000 year) performance of the proposed repository, from both chemical and radiological standpoints.

Environmental Restoration*

Over the 40 years of the Cold War with the Soviet Union, the primary mission of the Atomic Energy Commission, the Energy Research and Development Administration, and the Department of Energy was weapon production, with less attention to environmental controls. As a consequence, many sites around the country became contaminated with radioactive materials by the nature of the processes involved. Even before the need for nuclear weapons was reduced by events in the former U.S.S.R., it was seen that a major cleanup was necessary.

*Appreciation is extended to Victor S. Rezendes of the General Accounting Office for valuable information on this topic.

In 1988, Congress included language about environmental contamination in the Defense Authorization Act. Two descriptions of the needs were developed—the Glenn report and the Salgado report. In 1989 DOE Secretary James Watkins initiated a number of actions in response to evident needs. He stated that protection of the environment and assurance of safety and health took precedence over production, emphasizing the need for a new "safety culture." He established "Tiger Teams," consisting of DOE and contractor experts, which assessed compliance with rules and laws, and prepared reports that became bases for corrective action. The teams inspected 35 major facilities and issued 8715 findings. Subsequently, an Environmental Safety and Health (ES&H) Progress Assessment program was set up to implement the Tiger Teams' recommendations.

A key conclusion made by Congress, the National Research Council, the General Accounting Office, and DOE itself was that environmental and safety problems existed at many inactive sites and sites subject to closing. These sites contained stored hazardous and radioactive materials, and often had contaminated surfaces and equipment. A total of 111 U.S. inactive sites was identified, and a goal of cleanup by the year 2019 was set. To accomplish the 30-year task, DOE created the Office of Environmental Restoration and Waste Management (EM). Estimates vary greatly as to the cost of cleanup, ranging from $100 billion to $1 trillion, with the figure $300 billion frequently mentioned. Estimates also vary as to the number of facilities—buildings and areas—involved. The figure could go as high as 7000. The following table listing laboratories, sources of contamination, and estimated budget for 1995 provides some idea of the extent and nature of DOE's environmental restoration program.

The situation at Rocky Flats illustrates the problems to be encountered. This facility near Denver has been fabricating plutonium weapons parts for many years. In order to evaporate TRU-contaminated water, five "solar ponds" were dug. When concern arose about the effect on groundwater, a program of sediment removal was begun. Sludge was mixed with concrete to form blocks of what was called "pondcrete." These were discovered to actually be mixed waste (see Chapter 19). Moreover, blocks crumbled because of improper preparation. That waste will have to be repacked, and four more ponds will have to be cleaned out. Material classed as TRU will go to the WIPP site in New Mexico, and LLW will go to a disposal site in Nevada.

Many issues and challenges are facing the program of environmental restoration:

(a) Coordination with EPA Superfund program. The Environmental Protection Agency has identified 25 radioactively contaminated sites, including Maxey Flats, Hanford, and Rocky Flats, along with sites with soil containing natural radionuclides like uranium, thorium, and radium.

(b) The extent of cleanup that is feasible. A fundamental question is whether every site must be restored perfectly at an enormous

DOE Installations: Environmental Restoration and Waste Management
(Reference: Five-Year Plan FY 1994-1998)

Site	State	Source	FY 1995 Budget $ Million (est.)
Hanford	WA	Plutonium production, waste R&D	1736
Savannah River	SC	Tritium and Pu production	768
Fernald	OH	Uranium metals and compounds	373
Idaho Natl. Eng. Lab.	ID	Reactors, fuel processing	337
Rocky Flats	CO	Weapons component fabrication	334
Oak Ridge K-25	TN	Gaseous diffusion isotope separation	300
WIPP*	NM	Test repository for TRU	226
Oak Ridge X-10	TN	Energy technologies	226
Los Alamos	NM	Nuclear weapons, reactors	204
Oak Ridge Y-12	TN	Isotope separation, weapons materials	157
West Valley	NY	Reprocessing, waste solidification	134
Mill tailings	Var.	24 uranium mill tailing sites in West	89
FUSRAP†	Var.	28 Manhattan Project sites	76
Lawrence Livermore Natl. Lab.	CA	Explosives, hazardous wastes	68
Nevada Test Site	NV	Nuclear weapons tests	65
Sandia Natl. Lab.	NM	Non-nuclear components of weapons	54
Weldon Springs	MO	Processing uranium and thorium	54
Mound	OH	Weapons components	54
Pantex	TX	Weapons assembly, disassembly, storage	45
Portsmouth	OH	Gaseous diffusion isotope separation	33
Argonne Natl. Lab.-East	IL	Basic and applied research, reactors	29
Grand Junction	CO	Uranium mill tailings	28
Paducah	KY	Gaseous diffusion isotope separation	27
Kansas City	MO	Nonradioactive weapons parts	19
Brookhaven Natl. Lab.	NY	Research and development in science	18
Three Laboratories	Var.	Misc. (Ames, Fermi, Princeton)	16
Battelle	OH	General research for DOE	15
Lawrence Berkeley	CA	Research, accelerators	11
Santa Susana	CA	Reactor, fabrication	10
Pinellas	FL	Special weapons components	9
Sandia Livermore	CA	Non-nuclear components of weapons	7
U. California Davis	CA	Research on radiation effects	6
Other	Var.	Five other laboratories	18
Total			5546

*Waste Isolation Pilot Plant
†Formerly Utilized Sites Remedial Action Program

cost, in view of other environmental and societal needs, and in light of the national debt and budget deficits.

(c)   The need for new technologies. Examples are robots capable of decontaminating buildings and equipment, minimizing the exposure of humans to radiation; capability to process mixed waste; and techniques for remediation of contaminated groundwater. New technologies can save money, but it is uncertain whether a given method will be ready when needed and will be successful.

(d)   Requirements for environmentally trained personnel. The number of people with the necessary education and experience is limited, and it takes time to retrain and deploy people such as those formerly dedicated to weapons work.

(e)   Achieving understanding and support by the public for a long and expensive operation, in which there are few visible products for the funds spent. The public must also be satisfied that the cleanup is adequate.

(f)   Establishing priorities for cleanup. Much information about the facilities' contents is needed to make judgments. It is well-known that there is much uranium and plutonium, but there are also PCBs, hazardous asbestos, and lithium to deal with. Some of the stored enriched uranium is in an exotic form such as fluorides, as used in an experimental reactor. Some facilities can easily be converted to other use, some are too specialized for any future application, and some would have to be completely destroyed.

(g)   Deciding on the applicability of the concepts of ALARA (as low as reasonably achievable, see Chapter 6) and BRC (below regulatory concern, see Chapter 19).

(h)   Finding sites for disposal facilities that will accommodate the waste produced by decontamination and decommissioning.

(i)   Need to manage effectively. In view of the magnitude and complexity of the task, a comprehensive management system is needed. This involves thorough oversight of contractors instead of the "hands-off" policy of the past. It is regarded as necessary to build flexibility into contracting, in view of inevitable uncertainties.

The end of the Cold War has had several effects on the U.S. First is considerable relief after a half century of concern about nuclear war. Second is the elimination of need for massive defense expenditures, with the attractive possibility of a "peace dividend." It is realized, however, that the cleanup will involve a length of time comparable to that of the Cold War, and at a sizeable fraction of the cost. Another effect is the potential loss of jobs in the defense industry, including weapons production centers and national laboratories. Opportunities are available, however, to take advantage of the situation. One is the utilization of skilled and experienced defense personnel to achieve needed waste management and environmental restoration. Another is the preparation of DOE facilities for safe transfer to unrestricted public use, with emphasis on economic development in the affected geographic areas. New R&D and manufacturing enterprises can substitute for the defense programs of the past. There are indications that citizens of a region near a nuclear facility like Hanford or Rocky Flats are receptive to such trends because of their familiarity with radiation and their desire to retain community stability.

Disposition of Nuclear Weapons Material

The dramatic change in relations between the United States and the republics of the former U.S.S.R. has an indirect impact on the waste situation. The signing of the Strategic Arms Reduction Treaties (START) will result in large surplus stockpiles of highly enriched uranium (HEU) and weapons-grade plutonium.

Arrangements have been made for the U.S. to purchase HEU from Russia, for dilution in both countries to typical low-enrichment uranium (LEU) by the addition of natural or depleted uranium. The LEU will then be used as fuel for nuclear power plants. It is planned to process HEU at a rate of 10,000 kg per year for 5 years and then at 30,000 kg per year for 15 years. Some additional wastes will be generated in the conversion facilities. Their nature will depend on the blending process.

The fate of the plutonium, which is vulnerable to diversion, is yet to be determined. The disassembly of several thousand U.S. weapons will take place at the Pantex Plant near Amarillo, Texas. This DOE facility has been used for making the chemical high explosive for nuclear weapons, and for assembling and storing weapons. Warheads and missiles will be shipped from military forces to Pantex in special armored vehicles with escort and radio tracking. Disassembly will be done in cells called "Gravel Gerties," so named for the mass of gravel to absorb energy should a chemical explosion occur. The high explosive will be carefully removed from the plutonium "pits," which are stored in special bunkers called "igloos," with massive concrete blocks placed in front of their doors.

Several options for the disposition of the excess plutonium are available. One is to continue secure storage, with rigorous accountability to avoid diversion. The World Trade Center bombing heightened concern for the possible use of weapons material in threats or action. There is similar concern about theft for sale to a country seeking nuclear weapon capability. Another option is to convert the plutonium to oxide and blend it with natural or depleted uranium oxide to form mixed oxide (MOX) as a nuclear fuel. One serious proposal by an industrial consortium for such operations is called Isaiah Project, after the biblical phrase (Isaiah 2:4), "...beat their swords into plowshares and their spears into pruning hooks." It would involve bringing two partially completed but mothballed Washington State power reactors into operation. The burning of plutonium in this manner removes it from circulation while obtaining electric power. The spent fuel would contain a higher content of plutonium-240, whose spontaneous neutron emission makes the mixture unsuitable for weapons use. Another choice is to save the plutonium for initial loading of future breeder reactors. Finally, it has been suggested that the plutonium be denatured by combining it with high-level defense waste, or "spiked" with radioactive cesium-137 or a neutron-absorbing chemical such as gadolinium that is difficult to remove. The National Security Council will recommend which option to adopt.

A comprehensive study of the options and issues will reveal strong differences in opinion, related to concerns about proliferation of nuclear weapons, the desire to derive economic benefit, and attitudes about nuclear energy in general.

CHAPTER 21

Disposal of Spent Fuel

The rate of accumulation of spent fuel as a by-product of nuclear power generation is easily predicted. Most of the reactors on order in the U.S. have been completed and are operating. Whenever possible, those reactors will operate at maximum capacity in the interest of economy. The amount of uranium-235 consumed depends largely on the energy produced, and the requirement to remove fuel is based on the consumption of fissile material. Thus, assuming a once-through cycle, one can estimate rather accurately how much spent fuel must be managed in the next few decades. The graph shows this amount up to the year 2030.

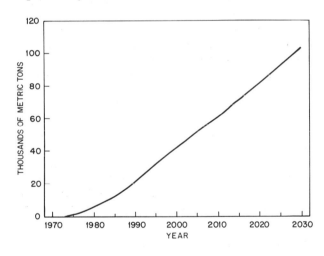

Total commercial spent fuel discharges, 1970-2030. (Adapted from Department of Energy Report DOE/RW-0006 Rev. 8.)

What to do with spent fuel has been an important challenge for the U.S. since the early 1970s and will continue to be well into the next century. To answer that question, several others must be posed:
- What are technically possible options?
- What is the presently economical strategy?
- What long-term alternatives are there?

One can readily think of many ways to manage spent fuel and other high-level wastes:
(a) Dilute and disperse in the air and large bodies of water. This is an unacceptable method for evident environmental reasons.

(b)   Delay action. This approach is favored by those who believe that the present political climate makes disposal anywhere impossible and who expect public opinion about radioactive waste to become less negative in the future.

(c)   Store for future use. Many in the nuclear community feel that as conventional energy sources are depleted in the 21st century, materials in spent fuel will be needed for breeder reactors.

(d)   Reprocess, retrieve, and recycle. Fissionable elements are especially useful, while some of the radioactive fission products have economic value.

(e)   Isolate spent fuel. This is the current preferred method to prevent radioactivity and radiation from affecting living beings and the environment.

In the following paragraphs, we will explore the various processes for obtaining options (d) and (e).

Reprocessing and Fractionation

In the early days of nuclear power, it was planned to reprocess spent fuel to separate fission products from uranium and plutonium, which would be recycled. Circumstances described in Chapter 14 currently prevent that option from being exercised in the U.S. The economic benefit of reprocessing is marginal at best. No large-scale reprocessing facilities are available. More importantly, in the existing political and economic uncertainty set by public opinion, the industry has little enthusiasm for developing, licensing, and constructing a reprocessing plant.

It is of interest, however, to examine some benefits of reprocessing and of fractionation, which is another word for "separating." Separation to prepare for future treatment or use is called "partitioning." First, let us look at the composition of residues from reprocessing. For each metric ton (1000 kg) of spent fuel the weights in the high-level residue are as listed below. Note that the weight of reprocessed

Composition of Residues from Reprocessing

Material	Weight, kg
Fission products	28.8
Fuel	
Uranium	4.8
Plutonium	0.04
Transuranics	
Neptunium	0.48
Americium	0.14
Curium	0.04
Reprocessing chemicals	68.5
Total	102.8

waste material is one-tenth the weight of spent fuel. Of course, since most of the radioactivity is from fission products, the activity per gram of waste is considerably higher.

If the fuel is going to be chemically processed to isolate five elements, it would be prudent to go further and separate out additional groups or individual elements. Recall that strontium-90 and cesium-137 are the worst offending isotopes during the first few hundred years of waste storage. If they were extracted, as was done at Hanford, the residual wastes would contain only a thousandth as much activity. The volume of those radioisotopes is small, and storage could easily be managed. On the other hand, capsules of cesium-137 with its 0.66 MeV gamma ray would be very valuable to the future food irradiation industry as a substitute for cobalt-60, which must be made in a reactor and which uses up neutrons.

By means of existing chemical technology or an extension based on research, there is a possibility of isolating some of the radionuclides of very long half-life that dominate the performance of a repository. Examples are technetium-99 (213,000 years), neptunium-237 (2.14 million years), and iodine-129 (17 million years). Removal of both the shorter-lived heat producers and longer-lived hazardous isotopes might make the disposal process simpler, cheaper, safer, and more acceptable politically. These long-lived isotopes could be stored separately, with only minimum shielding required.

Objections can be raised to the operation of plants for reprocessing and fractionation of spent fuel, because it might make plutonium more accessible for illegal purposes, including the manufacture of nuclear weapons or its use in terrorism. In contrast, one can argue that the best thing to do with plutonium is to place it in a reactor core, which is intensely radioactive already, and burn it there to help produce electric energy.

Another reason for not treating spent fuel is the increase in radiation exposure for workers required to operate, maintain, and repair a reprocessing plant. There is, however, a good possibility of developing remotely maintained facilities, especially if new electronic monitoring and control equipment is coupled to robotic technology. There is an analogy with future space exploration, in which human beings may not be exposed to risk if the missions can be accomplished by mechanical and electrical devices instead.

Transmutation

The word "transmute" means to convert one element into another by nuclear reaction. It is like the goal of the alchemist of the Middle Ages—to turn base metal into gold. Such a process is now possible through neutron bombardment, at a great cost, of course.

As applied to nuclear wastes, transmutation would involve irradiation of wastes by neutrons as in a fission reactor or in a future fusion reactor. The neutrons are absorbed to produce new isotopes that may have a short half-life or be stable. The process aids natural decay by effectively shortening half-lives.

The elimination of cesium and strontium is difficult because of the high neutron streams required. More likely targets are the transuranic elements plutonium, neptunium, americium, curium, and californium. Other candidates for transmutation are iodine-129 and technetium-99. The elements would be partitioned in a special reprocessing system, fabricated into fuel, and used in a reactor. Because some nuclei are fissile, they would yield power. Disadvantages include the possibility of creating heavy isotopes that decay with undesirable isotopes in the chain. Extra protection would be required in fuel fabrication because of neutron emission and residual radioactivity.

The method appears to be very attractive, but it is currently not feasible on the large scale needed. If research and development is continued on the Integral Fast Reactor (Chapter 14), burning of transuranic radionuclides may become a reality. The on-site fuel reprocessing facility that accompanies the reactor would make it safe and convenient to recycle certain isotopes.

Disposal in Space

Achievements of the U.S. space program suggest the possibility of disposing of wastes outside the earth using rockets. Several ways of handling containers of waste can be imagined. In order of increasing difficulty they are: placing them in orbit about our planet; putting them in orbit about the moon or depositing them on its surface; sending them into orbit about the sun as new "asteroids"; sending them into deep space; and shooting them into the sun. Rather than sending all wastes into space, it would be better to send partitioned materials of very long half-life. Neutron emission might pose a special shielding problem.

Space disposal would seem to be an ideal solution for getting rid of wastes permanently, but there are some drawbacks: (a) the possibility of an aborted mission, with the spacecraft burning up on re-entry, leading to atmospheric contamination; (b) the high cost of the vehicle to carry the propellant needed for extra shielding weight; and (c) reluctance to contaminate some extraterrestrial culture or future human habitat. Engineering studies have been made of the concept illustrated—to use the shuttle to take waste containers from the ground to a low orbit about the earth, then transfer them to an orbit about the sun, halfway between Earth and Venus. Although the method was considered feasible, costs were prohibitive.

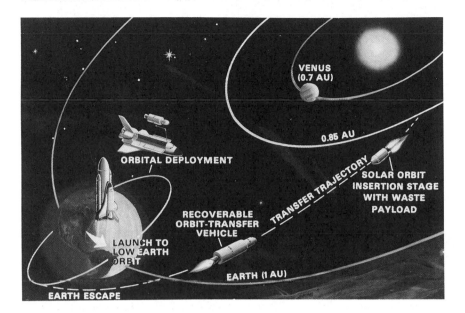

Extraterrestrial disposal of wastes by use of rockets. In this concept, the *Columbia* shuttle takes wastes from earth to orbit around the earth, then they are put in orbit around the sun. (Courtesy of the Marshall Space Flight Center, National Aeronautics and Space Administration.)

Ice-Sheet Disposal

Intuition tells us that the farther wastes are removed from habitation, the safer people will be. One of the most remote sites on earth is the polar ice cap in Antarctica, and several methods of disposing of waste there have been visualized, as shown in the sketch. In the most intriguing proposal, containers would be allowed to melt down through the ice by means of their own decay heat, with melted ice freezing above them as they descend. The canisters would eventually sink to the solid rock base, a mile or so

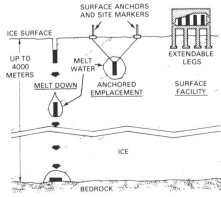

Ice-sheet disposal of solidified radioactive wastes. In this scheme, containers are supported at the surface or allowed to melt their way down to bedrock. (From *High-Level Radioactive Waste Management Alternatives.* WASH-1297, U.S. Atomic Energy Commission, 1974.)

down. This would appear to be a very secure disposal, but there is possibly a water layer between the ice and the rock. The layer would be produced by the great weight of ice, analogous to the film of water that allows a skater to glide over ice. Thus canisters would be exposed to water that is connected to the sea.

The ice disposal method is unacceptable because of the complexities of international ownership of Antarctica, the expense of long-distance transport, and the very brief time window allowed by the weather for entry to a site and emplacement of wastes.

Seabed Disposal

Low-level wastes have been dumped at various times into the water of the ocean, either as liquid or in concrete containers. However, the U.S. stopped ocean disposal in 1970.

Other techniques involving the sea are shown in the simplified sketch. Holes are drilled into the seabed from a ship, waste canisters are inserted, and plugs of inert material added. Sites chosen would be free of water currents and seismic activity. The method is similar to offshore drilling. In a variation of the idea, canisters are fitted with a sharp point and fins and allowed to drop like arrows in the water, burying themselves in the bottom sediment. It would take many years for the radioactivity to diffuse to the sediment surface.

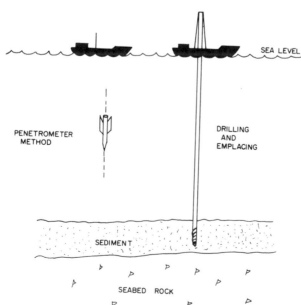

The seabed disposal technique. Holes are drilled in the ocean floor, or pointed canisters are allowed to bury themselves deep in the sediment at the bottom. Seabed disposal is considered a backup option. (From Raymond L. Murray, "Radioactive Waste Storage and Disposal," *Proceedings of the IEEE*, Vol. 74, No. 4, April 1986, p. 552 ff.)

Although there are different opinions as to which of the two approaches is better, seabed disposal is attractive. There is an enormous area of ocean floor that is far from civilization, has few life forms, and no minerals that explorers would seek. Even if containers failed, there is an enormous volume of water for dilution. Seabed disposal is viewed as a backup for more conventional disposal in the ground.

Geologic Disposal

Geologic disposal simply means burial in the earth. Several ways have been considered. First is placement of wastes in a very deep hole, going down 6 to 10 miles beneath the surface, and well below the level of moving groundwater. Canisters would be lowered into the hole and stacked in a column some miles in height, with the hole plugged. The advantage is remoteness; the disadvantage is lack of experience with large holes at such depths. Also, it is difficult to learn geologic features at such depths.

Second is a rock-melting technique, in which wastes would be dropped into a hole a mile deep. As visualized, the heat from radioactive decay melts the rock at the bottom, the wastes mix with the rock, and the mass of rock-waste eventually cools and solidifies, after perhaps as long as a thousand years.

Third is a process called "hydrofracture," in which water is forced into rock such as shale, causing layers to separate. Then waste mixed with cement or clay is pumped into the spaces, where it solidifies. The method has been used in the U.S. and the former U.S.S.R., but it requires very complete knowledge of the geology.

Fourth is the island isolation method. Waste containers would be placed below the fresh water table on an uninhabited small island far from civilization. The concept lies between seabed disposal and land disposal. It is not high on the priority list.

Fifth, and the preferred geologic method, is use of a mined cavity. As shown in the cutaway view, a shaft would extend from the earth's surface down to a series of horizontal tunnels in the geologic medium. Canisters containing spent fuel would be placed in holes drilled in the tunnel floor, as shown in the artist's drawing. Then the openings would be backfilled. An

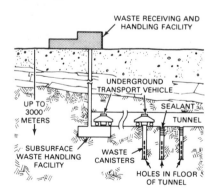

Emplacement of waste canisters in a mined cavity, a currently favored method. (Adapted from *High-Level Radioactive Waste Management Alternatives.* WASH-1297, U.S. Atomic Energy Commission, 1974.)

advantage is that conventional mining techniques can be applied, with some ability to correct defects if necessary. A disadvantage shared by several methods is the possibility of heat and radiation effects on the medium. Decay over the long storage period before disposal will alleviate that problem somewhat.

Artist's drawing of placement of canisters of waste in holes in the mine tunnel floor.
(Courtesy of Battelle Memorial Institute.)

Choices of Geologic Medium

Several materials have been promising candidates for the host geologic medium for a U.S. high-level waste repository:
(a) Rock salt, as recommended by scientific bodies and as used in the WIPP site (Chapter 20). In addition to New Mexico, salt is found in tall underground columns called salt domes in Texas, Louisiana, and Mississippi.
(b) Granite, prevalent along the Appalachian chain and the Great Lakes region. Earlier plans for a site in the East had considered this form of crystalline rock.
(c) Basalt, a rock formed by lava from volcanoes, found in the Northwest near the Columbia River.
(d) Tuff, a compacted and hardened ash from volcanic action millions of years ago. It is found in the Southwestern U.S., in particular at Yucca Mountain in Nevada, the possible site of the future high-level waste repository.

The ability of these geological media to greatly retard the migration of wastes in water, allowing decay to occur, depends on physical and chemical properties. One is porosity, which is the fraction that is space. Pore size and connection are important: small pores only slightly connected retard migration best. Second is permeability, a number that is related to ease of water flow. We list its value for several materials, all with a porosity of about one-half:

Permeabilities of Geologic Media

Material	Permeability
Clay	0.000016
Silt	0.33
Sand	25.3
Gravel	1130

Roughly, these numbers represent the rate of water flow under a certain water pressure. If the rock is cracked, the flow may be higher than if it is uniform.

The interaction between the radioactive contaminant and the geologic medium is expressed by a quantity labeled K_d, called the distribution coefficient of the chemical reaction. If it is large, the waste moves very slowly; if it is small, the waste moves with the same speed as the water. The retardation factor R, which is the ratio of water speed to contaminant speed, depends on K_d and on density ρ and porosity ϕ through the formula $R = 1 + K_d \, \rho/\phi$. Note that R goes to 1 for a very porous medium and increases with the reaction constant. Approximate values of R for selected elements and media are listed as follows:

Retardation of Elements in Geologic Media
(From *A Study of the Isolation System for Geological Disposal of Radioactive Wastes*,
National Academy Press, Washington, D.C. 1983, page 147.)

Element	Medium	Retardation Factor
Iodine	All	1
Technetium	All	5
Cesium	Salt	10
Neptunium	Tuff	100
Plutonium	Basalt	500
Thorium	Granite	5000

The wide variation of R with element means that each radioisotope must be considered separately in the analysis and calculation of system performance.

Characterization of a Repository

A high-level radioactive waste repository must protect people and the environment for many centuries. Any transport of material must result in a radiation dose that is less than a regulatory limit such as 25 millirems per year. To ensure this objective, the system, consisting of waste form, containers, backfill, and geologic medium, must minimize water access and contaminant escape. The waste must be emplaced in a setting that is safe initially and is likely to remain so for a very long time. Reference is frequently made to 10,000 years as the approximate time for the waste to be comparable in hazard to that of unmined uranium ore.

To achieve confidence in the security of the waste, comprehensive site characterization studies are made. The investigation is accompanied by analyses and computations that predict the radiation exposure over time. A brief description of the action plans for the Yucca Mountain Project (YMP) will give us perspective on the magnitude and complexity of a full characterization process. The plans are designed to lead to a decision as to whether the site is suitable. If it is, commercial spent fuel and vitrified defense wastes will be emplaced there.

Yucca Mountain, seen in the photograph, is in Nevada, about 100 miles northwest of Las Vegas, and very near the Nevada Test Site for nuclear weapons. The water table is about 1800 feet below the surface, and the average yearly rainfall is only six inches. Thus very little water gets down to the 1000-foot level where the waste would be located. It is believed that heat from decaying fission products would dry the rock surrounding the waste containers. Most of the required studies relate to the unsaturated zone, above the water table.

Yucca Mountain, the potential site of a waste repository. This aerial view of the northeast shows the Exploratory Studies Facility. (Courtesy of the Yucca Mountain Site Characterization Project Office, U.S. Department of Energy.)

Many sciences, popularly called "ologies," are applied: geology—the nature of rock; hydrology—the flow of water; seismology—earthquake effects; volcanology—potential for eruption of volcanoes; meteorology—climate and weather patterns; ecology—the science of biological systems—as well as law, sociology, demography, and politics. Some of the sciences, such as geology and hydrology, are closely related to each other; for example, fractures in rock affect the rate of water flow. Involved in the Yucca Mountain investigation are hundreds of scientists from many organizations. Included are the Department of Energy, the U.S. Geological Survey, and the national laboratories—Sandia, Los Alamos, and Lawrence Livermore—as well as several companies.

The testing program consists of drilling deep holes to find out how the geology varies in three directions; cutting trenches to investigate geologic faults, in which changes could cause earthquakes; measuring water presence, chemical composition, and flow patterns; and determining the presence of valuable mineral deposits that might attract intruders. To focus on the needs for information, the YMP has set up an Exploratory Studies Facility. This is an underground laboratory consisting of ramps down from the surface to a series of tunnels. A number of boreholes have been drilled using a minimum of water and using dust suppression systems to avoid air pollution. Several trenches have been dug to look for geologic faults and to make predictions of seismic activity.

The above physical studies are supplemented by observations and data collection on environmental and societal aspects of the region. Animal and plant species are identified and tabulated, with special attention to those classified as protected, threatened, or endangered. Desert tortoises, for instance, are rare and threatened, while lizards, ground squirrels, and scarab beetles are common and unprotected. Information on factors that would affect the small human population in the area is also gathered.

The characterization process is expected to take about ten years to complete. The ultimate objective is to assess the suitability of Yucca Mountain as the site for a repository. A large number of individual qualifying and disqualifying criteria must be applied. A few examples would be:

Qualifying: environmental quality can be protected
               adequate transportation routes are available
               societal effects can be minimized
               land acquisition is possible
Disqualifying: population density nearby is high
               site is in a federally protected area
               significant pathways from previous mining are
               present.

In the next two chapters, we discuss the laws and regulations that govern waste management, and the societal aspects of the disposal of all types of radioactive waste.

CHAPTER 22

Laws, Regulations, and Programs

Laws and Regulations on Atomic Energy

The management of radioactive materials has been subject to laws and regulations for over 40 years. Requirements have evolved as experience was gained, and programs have changed with the political climate. To review the historical background will help us understand the current situation.

The Atomic Energy Act of 1946 established the Atomic Energy Commission (AEC) and directed it to conduct research and development on peaceful applications of fissionable and radioactive materials. The Atomic Energy Act of 1954 emphasized both domestic and international uses of the atom. It also provided for the control of "source material" (uranium and thorium), "special nuclear material" (plutonium and enriched uranium), and "by-product material" (radioactive substances).

The National Environmental Policy Act of 1969 (NEPA) has as its purpose "to prevent or eliminate damage to the environment and biosphere and stimulate the health and welfare of man." The Act created the Council on Environmental Quality (CEQ), an advisory and coordinating group reporting to the President. NEPA requires every federal action that may significantly affect the environment to be accompanied by an environmental impact statement (EIS). An EIS is a large document that describes alternatives, potential environmental, social and economic effects, includes public comments and agency answers; and reports the findings of hearing boards. The Environmental Protection Agency (EPA) was created, with responsibility for air and water standards, limits on pollutants, and control of radioactivity. The EPA provides for public participation through meetings, hearings, and advisory group reviews.

The Energy Reorganization Act of 1974 divided the function of the AEC—developmental and regulatory—between two agencies, the Energy Research and Development Administration (ERDA) and the Nuclear Regulatory Commission (NRC). The Arab oil boycott of 1974 caused the U.S. to restructure its energy plans. The Energy Organization Act of 1977 replaced ERDA by the current Department of Energy (DOE), which also absorbed several other energy-related government programs.

The NRC has jurisdiction over reactor construction and operation. It also licenses and regulates the possession, use, transportation, handling, and disposal of radioactive materials, including wastes. Within NRC guidelines Agreement States accept authority to control radioactive materials. The principal reference is the *Code of Federal Regulations Title 10 Energy*. For example, 10 CFR 20 covers radiation standards and 10 CFR 50 covers reactors. Most people do not realize how highly regulated nuclear energy activities are. Title 10 Energy contains about 1400 pages of rules. The recurring theme is assurance of health and safety of the public and protection of the environment.

Other important agencies are the Department of Transportation (DOT), which provides rules on the shipment of all kinds of radioactive materials, and the Federal Emergency Management Agency (FEMA), which has plans for response to emergencies that involve the release of radioactivity.

The relationship of the three agencies that most directly affect waste management can be summarized as follows: EPA provides radiation protection standards; NRC licenses and regulates waste disposal under EPA limits; DOE does research and development, advises and helps with low-level waste disposal programs, and will operate defense and high-level commercial waste repositories.

Major legislation in the areas of LLW (1980, 1985), HLW (1982, 1987), TRU (1992), and energy (1992) provides the basis for the U.S. programs in waste disposal. The sequence of laws reflects responses to problems and to changes in public opinion. Regulations implementing the laws continue to evolve, affected by court decisions. There is a complex interaction among Congress, the States, regulating agencies, industry, and the judicial system. We briefly review the situation in the following sections.

The Low-Level Waste Policy Acts

Concern about the possibility of having no place to dispose of low-level radioactive wastes led the states to seek control of waste management. This led to the enactment in 1980 of the Low-Level Radioactive Waste Policy Act. It says, "Each state is responsible for providing for the availability of capacity either within or outside the state for disposal of low-level radioactive waste generated within its borders (excluding defense or other federal wastes). Low-level radioactive waste can be most safely and efficiently managed on a regional basis." The Act goes on to say that the states may enter into compacts with their neighbors under Congressional authorization. The law called for the ability of compacts to exclude wastes from other regions after January 1, 1986. The legislation led to the formation of a number of compacts, including the Appalachian, Central Midwest, Central States, Midwest, Northeast, Northwest, Rocky Mountain, Southeast,

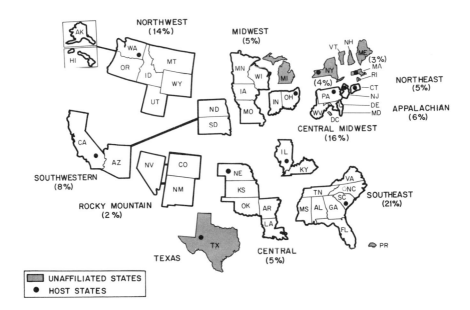

Interstate compacts for low-level waste. The host states are indicated by a dot. Maine and Vermont will join Texas to form a compact. Several states are unaffiliated. (Adapted from maps of the Nuclear Regulatory Commission and the General Accounting Office.)

and Western. Compacts are responsible for deciding what facilities are needed and which state will serve as host and for how long.

Negotiations among many states to form compacts and start developing disposal sites took longer than expected, making it impossible to meet the 1986 deadline. To relieve this problem and at the same time to expedite action, Congress passed the Low-Level Radioactive Waste Policy Amendments Act of 1985 (LLRWPAA). It called for keeping the three commercial disposal sites open for use by all through 1992. Annual and total limits on the volume of waste that can be sent from reactors were specified. LLRWPAA called for milestones and deadlines on ratifying compacts, selecting host states, developing plans, submitting license applications, and providing for disposal. Existing disposal sites were allowed to impose surcharges for disposal of wastes from regions without sites, with rebates to be used by states or compacts for site development. The Department of Energy kept track of these arrangements, with authority to allocate additional emergency disposal capacity to reactors, while the Nuclear Regulatory Commission could authorize emergency access to existing sites. The law also says that DOE is responsible for disposal of commercial LLW that is greater than Class C limits. DOE is required to assist states and compacts to achieve disposal facilities. DOE has a

National Low-Level Waste Management Program, which carries out studies, prepares reports, and provides technical assistance. It also administers the Host State Technical Coordinating Committee (TCC), composed of representatives from the dozen or so states or compacts that are developing low-level radioactive waste disposal facilities, with resource persons from federal agencies.

Progress toward establishing low-level waste disposal facilities by 1996 has been slow. The task has turned out to be much more complex and time-consuming than ever anticipated. A number of reasons for the delay can be cited: (a) each project must develop its own criteria and procedures for site selection; (b) a survey of a complete host state for potential sites involves the collection and analysis of large amounts of data; (c) the processes of site characterization and interpretation of data are long and involved; (d) the application for a license is a many-volume document; (e) the regulatory review of the application is thorough and extensive; and (f) lawsuits initiated by potential host communities or intervening groups delay action, sometimes by years.

The lack of disposal facilities has forced generators to store waste on site at considerable expense. In some cases, when storage costs were prohibitive, the use of radioactive materials was abandoned, to the detriment of the organization and its beneficiaries.

A Supreme Court ruling invalidated the part of the 1985 Amendments Act stipulating that states must take title to wastes if a facility was not available by 1996. The rest of the law remained intact.

The Nuclear Waste Policy Acts

For the management of high-level radioactive waste and spent fuel, The Nuclear Waste Policy Act of 1982 was passed by Congress. It was a compromise among industry, government, and environmentalists, containing timetables for action by the Department of Energy leading to underground disposal of HLW.

The Act relates primarily to commercially generated materials but provides for the disposal of defense wastes upon Presidential approval. A Nuclear Waste Fund was set up to pay for disposal, with money coming from the waste generators, who in turn charge users of electricity. A fee of 1/10 cent per kilowatt-hour is assessed. This is to be compared with a typical cost to the consumer of 6 cents per kilowatt-hour.

Following the dictates of the law, DOE set up an Office of Civilian Radioactive Waste Management (OCRWM), with its Director reporting to the Secretary of Energy. Guidelines were issued for the process to select suitable sites for a repository, a Mission Plan was developed, and geological surveys were begun. Nine sites were identified as potentially acceptable for the first repository, to be located in the West: in basalt at Hanford, Washington; bedded salt in the Paradox

Basin in Utah and the Palo Duro Basin in Texas; salt domes in Mississippi and Louisiana; and tuff at Yucca Mountain in southern Nevada. At the same time, possible locations were investigated for a second repository, in the East, in crystalline rock. The Great Lakes area and the Appalachian range were principal candidate regions.

The use of a Monitored Retrievable Storage (MRS) facility was studied, in accord with the law. Congress had visualized the MRS as an alternative to a repository, but the Civilian Waste Management Program conceived it as one part of an integrated disposal system. Fuel assemblies would be shipped from reactor storage pools to the MRS as a staging area, where fuel would be packaged and shipped to the disposal sites. The MRS could also provide backup storage capacity in case the opening of a repository was delayed.

In 1987, the choice of sites was narrowed to Hanford, Washington; Yucca Mountain, Nevada; and Deaf Smith County, Texas; and site characterization studies were begun. The search for repository sites and the MRS created much concern among citizens and lawmakers. DOE's decision to suspend the search for an eastern site was questioned. After much political compromise, Congress passed the Nuclear Waste Policy Amendments Act of 1987, which restructured DOE's HLW program. The only western site to be characterized would be Yucca Mountain. Nevada would receive financial compensation and special consideration in federal research projects.

The status of the MRS was redefined. A study commission would evaluate the need for it, and limits on fuel storage capacity of the facility are set at 10,000 tons. The NRC must issue a repository construction license before the MRS can be built; the license would be the legal device that would prevent the MRS from becoming a permanent storage facility. A Nuclear Waste Negotiator position was created, with assignment to seek a volunteer host for the MRS. Some movement toward that goal has occurred, since the first Negotiator, David Leroy, had interested several Indian tribes in the possibility of hosting the facility. Grants for study have been made to several tribes.

The 1987 Act had a number of special features. A Nuclear Waste Review Board in the National Academy of Sciences is created; spent fuel must be shipped in NRC-approved packages, with state and local authorities notified of shipments; authority is given for continued study of the subseabed disposal option, a topic of interest to European countries. No further crystalline rock studies are allowed, and DOE is to submit in the period 2007 to 2010 a study on the need for a second repository. The decision to characterize only one site will save considerable expense unless the Nevada site is found unsuitable, requiring other locations to be considered. The redirection of DOE's HLW program involves the preparation of a new limited scope Mission Plan and the release of only one Site Characterization Plan.

Progress in the characterization of the Yucca Mountain site has been extremely slow, largely because of efforts by the State of Nevada to stop the project. After several years of litigation, DOE obtained approval to proceed and not to be held up by requirements for permits.

Questions and concerns about the suitability of the site have been raised. The principal issue is whether there are pathways for rapid transfer of radionuclides to the environment. Site studies will be directed toward answering that question.

Plans have been made to meet a new schedule, calling for acceptance of spent fuel from utilities in 1998 and for beginning waste disposal in 2010. The ability of DOE to start taking fuel depends on the existence of a Monitored Retrieval Storage facility. If that is not ready, DOE may have to place fuel at national laboratory facilities.

Establishing the repository itself depends on the success of the OCRWM to characterize the site efficiently and to avoid obstruction by Nevada through acceptance by its citizens. DOE is committed to seeking another site if Yucca Mountain is found not suitable. Another requirement for success is continuous adequate funding for the project. Although the Nuclear Waste Fund is accumulating, it is necessary for Congress to appropriate the money for operations.

Energy Policy Act of 1992

This legislation was long in coming and broad in its scope, emphasizing energy efficiency, research and development on conventional fuels, alternative fuels, and uranium enrichment. The little that was said about radioactive wastes was important, however. The law specified that (a) EPA would set standards for Yucca Mountain based on findings by the National Academy of Sciences on several specific issues related to radiological protection, (b) NRC would provide requirements and criteria based on EPA standards, assuming engineered barriers and long-term oversight of the repository by DOE, (c) DOE (with NRC and EPA) would report to Congress on the adequacy of plans for disposal of waste from future reactors, and (d) states would have authority over below-regulatory-concern (BRC) wastes, negating NRC policy. (The NRC had hoped to exclude wastes that led to doses less than 1 mrem/yr, allowing them to be disposed of as ordinary waste. After failure to achieve agreement among interested parties, the NRC planned to permit disposal of BRC on a case-by-case basis. It is likely that most BRC will be disposed as LLW.)

Regulations on Low-Level Radioactive Wastes

Rules on civilian radioactive wastes in general are provided by the Nuclear Regulatory Commission. They are based on research by the NRC and its contractors. A Draft Rule with much supporting information is issued for review by those who would be affected, including industry and the public. After refinement, the rule is published officially in the Federal Register and in the *Code of Federal Regulations Title 10 Energy*. The principal regulation for low-level wastes is Part 61, usually called 10 CFR 61.

There is a wide range of activities in the low-level wastes to deal with, from barely above background to those comparable to high-level wastes. The lowest category is "below regulatory concern" (BRC), which can be disposed of without regard to radioactivity. Next are Class A wastes, which require "minimum" precautions for disposal. This means no use of cardboard containers, a need for liquid waste to be solidified or mixed with an absorbent so there is no more than 1 percent liquid, no explosive or spontaneously combustible material, at limited pressure if gaseous, and treated if of biological origin. Class B wastes must meet minimum requirements but also have "stability." This means they must keep their size and shape despite effects on containers from soil weight, moisture, or radiation. Class C wastes should be isolated from a future "inadvertent intruder," a person who accidentally comes upon the waste while digging in the area after the site has been closed. The person may be drilling a well or excavating for a building or cultivating the land. The C waste should be buried more deeply to protect the intruder. Wastes that are more active than C cannot be given near-surface disposal. These "greater than Class C" (GTCC) wastes must be treated as HLW, to be disposed of by the Department of Energy.

The boundaries between classes of wastes depend both on the isotope's half-life and the specific activity in curies per cubic meter. For example, when the concentration of tritium (12.3 years) reaches 40 Ci/m^3, it becomes B waste; when that of cesium-137 (30 years) reaches 4600 Ci/m^3, it becomes GTCC; when iodine-129 (17 million years) reaches 0.008 Ci/m^3, it goes from A to C. The regulation 10 CFR 61 gives details for all radionuclides.

The waste disposal facility is licensed either by a state or by the NRC for use by a commercial operator. The site is selected from several candidates on land owned by the state or federal government. All pertinent facts about the geology, water flow patterns, and nearby population must be known. Operations are inspected to be sure that the wastes are properly managed, and if so the license is renewed. At the end of the useful period of the facility, 20 to 30 years, the site is closed. The license is transferred to the state or federal agency, which will continue to monitor the site for the period of institutional control, which is 100 years. Then the license is terminated and no further maintenance is needed, but the design should have assured protection for a period of 500 years.

Supplementing the regulations are NRC documents called Regulatory Guides (colloquially "Reg Guides"), providing information on such things as quality assurance, design bases, calculation methods, and the form for reporting. An example is No. 4.18, "Standard Format and Content of Environmental Reports for Near-Surface Disposal of Radioactive Waste," 1983.

Regulations on High-Level Radioactive Wastes

As noted in an earlier section, the disposal of HLW is controlled by the NRC, under EPA standards. The principal regulation is Part 60 of the *Code of Federal Regulations Title 10 Energy*, known as 10 CFR 60. Some of its important provisions are distilled from more than 30 pages of regulations, as follows:

1. The design and operation of the facility should not pose an unreasonable risk to the health and safety of the public. The radiation dose limit is a small fraction of that due to natural background.

2. A multiple barrier approach is to be used, including the waste form, the containers, and the host rock.

3. Performance objectives are set for both the components and the system.

4. A thorough site characterization study must be made, with features such as possible flooding regarded as sufficient to disqualify, and features such as great geologic stability and slow water movement regarded as favorable.

5. The repository should be located where there are no attractive resources, far from population centers, and under federal control. Good records and prominent markers are required.

6. High-level wastes are to be retrievable for up to 50 years from the start of operations.

7. The waste package must be designed to take account of all possible effects; it must be dry and chemically inert.

8. The wastes in the package should be secure for at least 300 years. Groundwater travel time from the repository to the source of public water should be at least 1000 years. The annual release of radionuclides must be less than a thousandth of a percent of the amount of the radioactivity that is present 1000 years after the repository is closed.

9. Predictions of safety must be made with conservative assumptions and by calculations that take account of uncertainties, using expert opinion.

The above rules are augmented by 10 CFR 960, which contains the Department of Energy's criteria on characterizing repositories, developed in response to the original Act of 1982. In view of the action by Congress in 1987 to limit study to Yucca Mountain, the regulations related to selection of several sites for characterization and on the recommendation of one for use are now irrelevant.

Environmental Standards

The technical basis of licensing and regulation of potentially harmful substances is normally provided by the Environmental Protection Agency (EPA). Rules appear in the *Code of Federal Regulations Title 40 Environment*. In the case of radiation protection, standards were delayed and NRC initiated the rules. Also, some EPA rules have been challenged by the courts, requiring new rulemaking. Thus some EPA rules are in place while others are being developed.

The approach to standards taken by the EPA is to look at the possible hazards under a variety of situations ("scenarios") involving different sources of radionuclides, their method of treatment, protective measures taken, and health consequences. The dollar cost of each case is also calculated.

The evolution of EPA rules is difficult to describe because of the interaction with Congress, other agencies, and the public. We give only a few highlights. In Title 40 of the *Code of Federal Regulations* there are three Parts in Radiation Protection Programs. The first of these is 40 CFR 190, referring to nuclear power operations and limits on doses and releases in the uranium fuel cycle. The second, and probably most important, is 40 CFR 191 on management and disposal of spent nuclear fuel, high-level, and transuranic wastes. EPA seeks consistency of its rules, and expects to adopt generally the following: (a) to change the regulatory period on repositories from 1000 years to 10,000 years; (b) to adopt the newer concept of an annual committed effective dose (see Chapter 6), with a limit of 15 mrem; (c) to specify a drinking water standard corresponding to 4 mrem/year for beta-gamma emitters; and (d) to put limits on the release of gaseous carbon-14. Some of the new rules are being questioned; for example, DOE notes that the CO_2 rules are difficult to meet without expensive extra containers for waste, and that there exist other much larger unregulated industrial releases. The third Part is 40 CFR 192, having to do with uranium and thorium mill tailings. An example is the limit of radium-226 to 5 picocuries per gram in the top layer of soil.

In locating disposal facilities, wetlands must be avoided. Wetlands are bodies of water or land saturated with water that support plant and animal life, excluding farmland. They may be filled in only with special permit. The EPA and the U.S. Army Corps of Engineers administer the rules of Section 404 of the 1972 Clean Water Act.

The protection of wildlife is also under the EPA. Included are rare, threatened, or endangered species of animals and plants. An example that has received much attention in the western U.S. is the desert tortoise.

Organizations

A number of organizations besides governmental agencies contribute information and opinion on waste disposal. Public interest and environmental groups include the Sierra Club, Radioactive Waste Campaign, Union of Concerned Scientists (UCS), the National Resources Defense Council (NRDC), and Friends of the Earth. The nuclear industry is represented by the Nuclear Power Oversight Committee (NPOC), the Nuclear Management and Resources Council (NUMARC), the Electric Power Research Institute (EPRI), and the Nuclear Energy Institute (NEI), formerly the U.S. Council for Energy Awareness (USCEA).

Professional societies that publish and sponsor conferences on wastes include the American Nuclear Society, American Society of Mechanical Engineers, American Physical Society, Materials Research Society, American Chemical Society, American Institute of Chemical Engineers, American Ceramic Society, Institute of Electrical and Electronics Engineers, and Institute of Nuclear Materials Management. Four periodicals that deal with waste are *Waste Management* (Pergamon Press), *Radioactive Waste Management and the Nuclear Fuel Cycle* (Harwood, London), *Radioactive Waste Exchange*, and *Radwaste* (ANS). The annual symposium on Waste Management held in Tucson, Arizona, publishes proceedings on recent problems and progress. Other important organizations are the American National Standards Institute, the National Academy of Sciences, the National Governors' Association, the National Conference of State Legislatures, and the Low-Level Radioactive Waste Forum. Worldwide waste management is addressed by the International Atomic Energy Agency in Vienna and the Organisation for Economic Co-operation and Development, headquartered in Paris.

Radiological Tests

Late in 1993 it was revealed that, between 1945 and 1971, government agencies had conducted a number of radiological tests on human beings. Examples cited in the news media* were: (a) releases of radioactivity to public areas as part of Cold War investigations of fallout and radiological weapons; (b) irradiation of prisoners with x-rays to determine doses that produce sterility; (c) injection of plutonium into terminally ill patients; and (d) use of radioactive tracers in food given to mentally retarded children for metabolism studies. The

---

*Two of the editorials are: Gina Kolata, "Some scientists horrified, others wait to judge radiation tests," *New York Times*, January 9, 1994, and Jessica Mathews, "Radiation's legacy; secrecy's shame," *Washington Post*, January 9, 1994.

data from the research has been in the literature and has been used to determine safe and harmful levels of radiation. Those tests involving military applications were carried out in secret. Many of the studies were conducted without the knowledge or approval of the subjects, and some involved exposures now considered dangerous.

An investigation of the details and circumstances of the radiation tests is promised by government officials. Questions that may be asked are:

- What were the actual radiation doses received by the subjects?
- How should current standards on human rights be applied to past actions?
- What role does lack of knowledge about radiation effects play in assigning blame?
- What compensation is due those exposed to unacceptable levels of radiation?
- Is it appropriate to suppress the use of data which has been obtained improperly?

CHAPTER 23

Societal Aspects of Radioactive Wastes

We have seen how physical sciences and life sciences are involved in the management of radioactive wastes. Technical areas applied to ensure safe treatment and disposal are physics, chemistry, mathematics, engineering, geology, hydrology, seismology, climatology, meteorology, biology, ecology, and medicine. Despite the knowledge and talent available in these subjects, it has been difficult over the years to arrive at firm decisions and to achieve positive action in waste disposal. Part of the problem stems from the greater attention paid to the products of nuclear processes rather than to the by-products. Another part has been the many changes in direction through different national administrations. But a significant part of the problem has been public opposition.

We shall describe the situation, noting some of the views and attitudes about wastes. Such analysis may help the reader's appreciation of the matter and set the stage for a presentation of current approaches to solving both the technical and societal problems of radioactive waste disposal. Many different opinions exist that reflect personal philosophies and social values. These in turn vary with individual background, knowledge, personal experiences, and professional affiliation. In a democratic society such differences are to be expected. Convictions seem to be stronger on this subject than on most issues, making it hard to achieve agreement. Let us now begin to examine the relationship of technical and societal aspects and to indicate possible means by which they may be brought into harmony.

Opposition to change has increased in recent years, along with a worldwide decline in acceptance of the authority of government and industry, and a demand for greater public involvement in decisions. Reactions are not limited to waste disposal by any means. Almost any new facility is viewed as a threat, whether it be a shopping center, low-income housing, a home for physically or mentally ill, a shelter for homeless, a recycling plant, a waste incinerator, or a landfill.

"NIMBY"

An almost universal reaction to a waste disposal facility (shown in the conceptual drawing) is, "Not in my back yard." This attitude is so

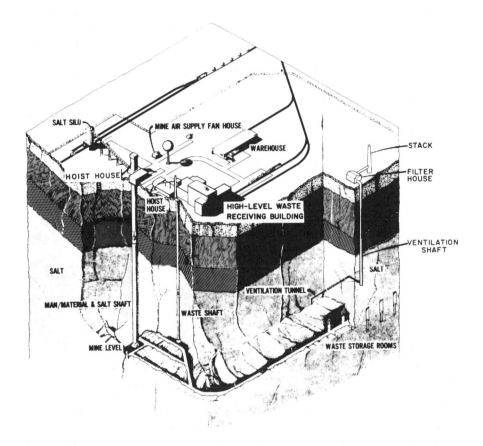

Artist's conception of a high-level radioactive waste repository. (Courtesy of Oak Ridge National Laboratory.)

prevalent that it is given the acronym "NIMBY." Several reasons give rise to the response, varying in intensity and ranging from the highly emotional to the very intellectual.

First is fear of the unknown among those with no scientific knowledge. Radiation is regarded as mysterious and hence dangerous. Even among informed people, there is fear of anything nuclear. A large fraction of the people living today have grown up under the threat of nuclear warfare, and they tend to associate bombs, nuclear reactors, radioactivity, radiation, and wastes. The potential for physical damage by nuclear weapons and the ability of radiation to produce cancer are well known and dominate many people's thinking, even in situations where the hazard is minimal.

Second is concern that a waste disposal facility will fail and cause long-term contamination of air and water, with consequent hazard to people and the environment. Causes cited are human error in design or operation, or merely the validity of "Murphy's Law," which says that if anything can go wrong, it will.* Some people may recognize that the potential danger is small but are concerned that they have no control over it, as they believe they have when driving an automobile, for example.

Third is the anticipation that the advent of a waste disposal facility will cause local land values to fall, will discourage new industry or tourism, or will change existing life styles. Such opinion can be modified by full understanding of benefits available to the region, including new tax and fee revenues and job opportunities.

Fourth is the opinion that waste disposal should be the responsibility of the power companies, since they are the principal generators. Believing that it is impossible to ensure isolation of wastes underground, many urge that wastes be stored above ground at nuclear stations. Long-term storage is said to allow new disposal options resulting from future research and development.

Fifth is distaste for being "imposed on" by having to accept a nuclear "dump," especially if those affected have had no voice in the decision. This reaction may be an extension of the frustration of observing local governments make changes in zoning, acquiescing to pressure from industry and developers. Federal government agencies such as the Department of Energy are perceived by many to be insensitive to public concerns.

The sixth is a philosophic position, related to the belief that nuclear power is symbolic of centralized authority. Some opponents hold that the waste problem would disappear if all nuclear plants were shut down or phased out rapidly. A corollary is that the power supply should be replaced by stricter conservation measures and the use of renewable power sources involving sunlight and wood, with authority widely distributed in the hands of the public.

The response by the public to a waste disposal project takes various forms: letters to the administrator of the program, to editors of newspapers, or to government officials; speeches against the project at public meetings; voting elected representatives out of office; holding protest rallies; initiating injunctions and lawsuits; and making threats of physical violence. It is never clear, however, how widespread the resistance actually is, for two reasons: (a) the intensity of opposing rhetoric can be very misleading as to what is the real public opinion; (b) many people who support the endeavor or are neutral are afraid to express themselves.

---

*Also see William and Mary Morris, *Morris Dictionary of Word and Phrase Origins*, Harper and Row, New York, 1972.

Public Participation

It is often noted that the solution to the nuclear waste problem is more social than technological. The narrowest interpretation of that statement is that it is necessary only to educate the public. Administrators who proceed solely on that assumption are doomed to failure. Four levels of interaction are needed among those who undertake any project and those affected. Providing people with information is necessary, but not sufficient. Next is allowing and seeking public comment. Still higher in importance is giving the public opportunity to influence decision through an agency policy that takes heed of people's opinions and of the additional information they supply. The final step would be to submit agency proposals to the public for approval. This procedure would have to be limited to major actions, not day-to-day operations. Some observers maintain that public approval is the only way to success, while others believe that the establishment of confidence and trust in the project leaders is the key. The use of prestigious academic bodies to review and comment on issues and plans has been employed with some favorable results. A supervising body composed of citizens is probably preferable to a government agency, except for the difficulty in finding knowledgeable people to serve who are not in some way associated with nuclear activities. If technical individuals are responsible for a waste disposal project, they must be convinced of the need for public understanding and public participation in establishing disposal facilities. They must be able and willing to devote considerable time and effort to communication and consultation with environmental and public-interest groups, advisory committees, and individual citizens.

Scientists and engineers involved in waste management find themselves in awkward positions when they encounter public opposition. When asked to certify safety, they must answer in terms of probability, which does little to relieve doubts. When attacked verbally as being incompetent or being unconcerned about human values, some may be resentful and defensive, resulting in further alienation of the public. Technically knowledgeable people who are most successful in working with the public are those who seek to understand the worries and fears of citizens. At the same time, scientists and engineers, who have been trained to speak the truth as best they know it, must be aware that some zealots are willing to say whatever is likely to further their cause. They also should realize that government agencies and industrial organizations tend to release no more information than forced to, or to release favorable information, to "maintain a low profile" against attack. These practices tend to cause the public to mistrust government and industry. To maintain credibility and enhance trust, technical people in waste management must always provide full, accurate, and reliable information.

Facts and Ideas

Information that can be provided the public to alleviate unwarranted concerns includes the following:

1. Wastes are an inevitable by-product of our society's technological basis, including uses of the nucleus for medical diagnosis and treatment, electric power, and defense.
2. Much research and development has been directed toward safety measures in waste treatment, container design, transportation, and geologic disposal.
3. Strict regulations are in place for handling radioactive materials and for designing and operating high-level waste repositories and low-level waste disposal sites.
4. Radiation exposure to the public from transportation of wastes and from any disposal facilities is a small fraction compared to that from natural radiation.
5. A suitable choice of geologic medium has the capability of filtering radioactive elements and preventing contamination of water.
6. Radioactive materials eventually decay, in contrast with many permanent hazardous chemicals.
7. The longer the half-life, the smaller is the amount of radiation released from radioactive substances, all other things being equal.
8. Radioactive materials are readily detected by their radiation, in contrast with difficulty in measuring traces of chemicals.
9. More is known about the biological effects of radiation than of any other hazard.
10. Continuing to store radioactive wastes rather than disposing of them would leave a problem for future generations.

Mutual Understanding

It is possible to visualize an ideal situation, in which radioactive wastes are managed to the satisfaction of both the responsible agency and the public. Appropriate philosophies, policies, and practices to help achieve success can be identified.

A fact to be accepted is that radioactive wastes exist and will continue to be generated for some time, whether or not any new nuclear plants are built.

The ultimate goal is the safe management of wastes, to minimize hazard to workers and the public. Zero risk cannot be guaranteed, but the risk is small compared with those normally accepted as part of human life.

Even under conditions of negligible risk to people and the environment, there can be some adverse effects of a waste facility, requiring that some financial incentives or other benefits to a community be provided.

The agency charged with the task of choosing a disposal site or of managing a facility will make every effort to provide accurate technical information about the project to all interested parties.

It is understood by all that participation by the public in projects that affect human beings and the environment has become an expected supplement to the classic representative government. Discussion and decisions on actions on waste disposal are to be carried out openly, with opportunity for anyone to attend meetings and contribute opinion, and to have access to working documents.

Interaction between citizens and agency representatives is to be characterized by mutual courtesy and respect in an atmosphere that seeks to avoid antagonism. Those having legal responsibility for actions involving waste disposal will be responsive to comments, questions, suggestions, and recommendations by individual members of the public or by organized groups, but will make the ultimate decisions on steps to be taken. Those charged with the task of managing radioactive wastes will not be content with being mere administrators but will also strive to be leaders who instill confidence through actions rather than words.

There is historic precedent for the relationship of people and leaders. We recall the pronouncement of the Declaration of Independence, ". . . governments are instituted among men, deriving their just powers from the consent of the governed. . . ."

CHAPTER 24
Perspectives—A Summing Up

We have sought to provide a brief but broad picture of radioactivity, radiation, nuclear power, radiation applications, and radioactive waste management. We have tried to be informative, factual, and fair. Let us now select some highlights, key ideas, and important conclusions about nuclear wastes.

1. Radioactivity is both natural and manmade. The decay process gives radiation in the form of alpha particles, beta particles, and gamma rays. All of us are exposed to natural background radiation coming from cosmic rays and minerals in the ground. Radon is an important source of exposure.

2. Radiation can be harmful to the body and to genes, but an effect at low levels cannot be proved. Several methods of protection are available, including time, distance, and shielding.

3. The fission process yields energy that is converted into electricity in nuclear plants, but also generates wastes in the form of radioactive fission products and activation products. It is difficult and expensive to destroy wastes.

4. There is a large volume of defense wastes, stemming from World War II and subsequent weapons production. Its disposal is yet to be completed, and a massive cleanup of facilities will be required. In response to changing needs, DOE has restructured to focus on dismantling nuclear weapons, developing renewable energy sources, transferring technologies to U.S. industry, and cleaning up former weapons production facilities.

5. Radioisotopes and radiation are used extensively in research, medical diagnosis and treatment, industry, and space applications. Many of the waste by-products are of short half-life.

6. Low-level radioactive waste comes from nuclear reactors, industry, and health-related institutions. Industry has greatly reduced waste volume over the years, but the number of waste burial sites available has dwindled. Programs to find new disposal sites are under way throughout the U.S. by states and regional compacts.

7. The most radioactive of the wastes to be disposed of is spent fuel from nuclear power reactors. It will accumulate at reactors in water pools or dry storage containers until a repository is built.

8. Attempts have been made for many years to find suitable locations for the disposal of high-level wastes, including used nuclear fuel. Emphasis on power production and frequent changes in program direction have delayed this goal.

9. The public is concerned about the transport of radioactive wastes, but isotopes, spent fuel, and other wastes are routinely transported in specially designed containers able to withstand conceivable accidents.

10. Research and development on the waste disposal process will continue. Plans for disposal set by Congress are to be carried out by the Department of Energy for spent fuel, and by the states for low-level waste. Decisions must be made about what is best to do with plutonium and highly enriched uranium produced by the U.S. and the former U.S.S.R. as part of the Cold War.

11. Any waste isolation system must protect the public and future generations from harm due to radiation exposure. Design goals are to isolate reactor wastes until they are no more dangerous than the ore from which nuclear fuel came, to keep the hazard as low as reasonably achievable, and to limit exposure to a small fraction of that due to normal background. Isolation need not be forever, but only until enough radioactive decay has taken place.

12. Waste management involves application of science and engineering, but also requires that the public participate in selecting disposal sites and technology. Adherence to regulations is vital to the success of any waste management project.

13. Several methods for disposing of spent fuel have been studied, with the conclusion that geologic disposal in a mined cavity deep in the earth is preferable. Disposal in the seabed is a backup approach, but unlikely to be used.

14. Interstate compacts have been formed in the U.S. to handle all aspects of low-level waste disposal—site selection, design, construction, operation, site closure, and long-term care, after which use is unrestricted.

15. The state of Nevada is scheduled to host the first high-level waste repository if extensive site studies show Yucca Mountain to be suitable. Completion will be early in the next century.

16. Protection of the public and the environment from radiation exposure by radioactive wastes is provided by multiple barriers which include waste form, containers, and the geologic medium.

17. For low-level wastes several alternatives to shallow land burial are available. Examples are intermediate-depth disposal, mined-cavity disposal, belowground vault, modular concrete canister disposal, earth-mounded concrete bunkers, shaft disposal, aboveground vault, and earth-mounded aboveground vault.

18. A legal structure exists to manage wastes safely, including laws specifying policy, requirements, and schedule; environmental protection standards; and licensing and regulation by the federal government and the states.

19. Laboratory and field experiments and tests yield data for use in calculation of predicted performance of disposal systems, but they cannot prove that there will be zero release of radioactivity. Technical experts believe that the hazard can be kept far below that acceptable from other sources of risk in daily living.

20. Nuclear processes have become a significant part of our economy and our culture through defense use, medical applications, and the generation of electric power. Long-term energy demands will be met only by use of all available sources, including nuclear. Wastes are an inevitable by-product and must be safely isolated.

21. Continued study of the phenomena of global warming, acid rain, and ozone depletion will tell what environmental benefit may derive from the use of nuclear power as a source of electricity.

22. It is a challenge to all of us to enhance the benefits and reduce the risks in the applications of the nucleus. This will involve dedicated and sincere efforts by program leaders to provide accurate and complete information, to understand and respond to concerns, and to establish meaningful dialog with those people who are affected.

GLOSSARY

Brief and simple definitions are given for words and phrases in this document and other literature on waste management. The four topics into which the glossary is divided are these:

Nuclear processes and radiation
Reactors and fuel
Waste characteristics
Geologic features.

Nuclear Processes and Radiation

Accelerator—an electric/magnetic device to give charged particles a very high speed.

Activation—producing radionuclides by irradiation, especially by neutrons.

Activity—rate of disintegration (also see **Curie**).

Alpha particle—a type of radiation; the helium nucleus.

Atomic number—number of electrons or protons in an atom.

Attenuation—reduction in radiation on passage through matter.

Becquerel (Bq)—unit of activity, 1 disintegration per second (d/s).

Beta particle—a type of radiation; the electron.

Biological half-life—time for half of a radioisotope to be eliminated from the body.

Curie (Ci)—a unit of radioactivity equal to 37 billion disintegrations per second (d/s).

Dating—finding the age of an object by its radioactivity.

Daughter—a nucleus that results from radioactive decay.

Decay—the disintegration process of nuclei.

Dose—amount of energy absorbed.

Electron—basic electrically-charged particle, mass 9.1×10^{-31} kilograms, charge 1.6×10^{-19} coulombs.

Electron volt (eV)—a unit of energy, 1.6×10^{-19} watt-seconds.

Fission—splitting of nuclei by neutrons.

Fission products—the nuclei, usually radioactive, resulting from fission.

Gamma ray—a type of radiation; a high energy photon or electromagnetic wave.

Genetic—an effect (as of radiation) on hereditary tissue.

Half-life—the length of time for half the atoms of a radioactive substance to decay.

Ionization—removal of electrons from an atom, for example, by means of radiation.

Isotope—atoms with the same atomic number but different mass number.

Joule—a unit of energy, the watt-second.

Linear hypothesis—the assumption that any radiation causes biological damage, according to a straight-line graph of health effect versus dose.

Mass number—number of neutrons plus protons in a nucleus.

Neutron—a basic particle that is electrically neutral, weighing nearly the same as the hydrogen atom.

Periodic table—a chart of the chemical elements.

Person-rem (or man-rem)—the product of average dose by the number of people affected.

Rad—a unit of radiation energy absorption; 1/100 joule per kilogram.

Radiation—particles or waves from atomic or nuclear processes (or from certain machines).

Radioactivity—spontaneous disintegration of an unstable nucleus.

Radioisotope—a radioactive isotope.

Radionuclide—a species of atom that is radioactive.

Rem—unit of radiation dosage equal to the rad for x-rays, gamma rays, and some beta particles; accounts for biological effect.

Somatic—a direct effect (as of radiation) on the health of tissue.

Tracer—an isotope used to follow a process.

X-rays—electromagnetic radiation of energy greater than that of visible light, usually produced by an x-ray machine.

Reactors and Fuel

Assembly—bundle of fuel rods used in a reactor.

Barrier—metal sieve used in gaseous diffusion isotope separation.

Boiling water reactor (BWR)—a light-water cooled reactor in which some boiling occurs.

Breeder reactor—a reactor that produces more fissile material than it consumes (by a process called "breeding").

Cladding—the outer coating of nuclear fuel, for example, a tube.

Converter—a reactor in which some fertile material is made into fissile material.

Criticality—a condition in which a chain reaction involving neutrons and fuel is self-sustaining.

Decommissioning—removal from service at the end of useful life.

Decontamination—the removal of radioactive material.

Enrichment—a process to increase the percentage of a desired isotope such as uranium-235.

Fabrication (of fuel assemblies)—making uranium oxide fuel pellets and forming fuel rods and bundles of rods.

Fertile—a material that becomes fissile upon absorbing a neutron.

Fissile—able to be split by a low-energy neutron.

Fuel—fissionable material "burned" in a nuclear reactor, for example, uranium.

Fuel cycle—all steps in supplying, using, and processing fuel for nuclear reactors, including disposal of wastes.

Fusion—a nuclear process in which nuclei are combined to yield energy.

Implosion—a compression to detonate a nuclear weapon.

Licensing—giving a permit to build or operate a facility.

Light-water reactor (LWR)—a nuclear reactor cooled and moderated by H_2O.

Moderator—a light element used to slow neutrons, as in a reactor.

Multiplication—neutron interaction with fissile material in a chain reaction.

Natural uranium—uranium as mined (0.7% ^{235}U, 99.3% ^{238}U).

Pile—an early name for nuclear reactor.

Plutonium—the element formed by neutron absorption in uranium-238.

Pressurized water reactor (PWR)—a light-water cooled reactor operated at high pressure without boiling.

Reactor—a device involving a chain reaction using neutrons.

Spent fuel—nuclear fuel that has been removed from a reactor after use to produce power.

Tonne—a metric ton, 1000 kilograms.

Waste Characteristics

Actinides—elements of the periodic table wth atomic number 90 through 103.

Barrier—a component that slows the movement of radioisotopes.

Calcine—powder produced by heat treatment.

Canister—the primary container for solid waste.

Ceramic—insoluble solid oxide.

Disposal—removal from man's environment permanently.

Grout—a cement mixed with wastes.

High-level wastes—fission products plus some actinides.

Ion exchange—a process used to purify chemicals.

Isolation—preventing migration of wastes to the biosphere.

Leaching—dissolving in a liquid.

Low-level wastes—those generally not requiring shielding or heat removal; small transuranic content.

Mill tailings—see **Tailings.**

Once-through—a fuel cycle in which spent fuel is not reprocessed.

Partitioning—separation of certain radioisotopes from waste.

Regulation—maintenance of standards of performance through rules.

Repository—a location for waste to be disposed of.

Reprocessing—the mechanical and chemical treatment of nuclear fuel to separate uranium, plutonium, and fission products.

Retrievable—able to reclaim if necessary.

Salt—sodium chloride, NaCl, as a geologic medium.

Storage—holding temporarily.

Tailings—the residue from extraction of uranium from its ore.

Transmutation—transformation of isotopes using nuclear reactions.

Transuranic—beyond uranium in the periodic table.

TRU—transuranic waste, with more than 10 nanocuries per gram.

Geologic Features

Anticline—see **Fold.**

Aquifer—underground layer of material through which water passes.

Bed—layered deposit of sediment in the form of rocks, products of weathering, organic materials, and precipitates. Also, **bedded.**

Biosphere—regions of the earth and atmosphere occupied by living beings.

Breccia—fragmented rock region (as in breccia pipe).

Diagenesis—the conversion of sediment into rock by compaction or chemical reaction.

Diapir—an anticline fold that has broken through the rocks above. Also, **diapirism.**

Dome—a bed that arches up to form a rounded peak deposit. Also, **domed.**

Fault—a break in a rock formation usually involving diagonal movement. An example: the San Andreas fault in California.

Fold—a curved deformation of rock. The peaks are called anticlines, the valleys, synclines.

Glacier—large body of ice, often moving slowly.

Igneous rocks—formed by solidification of molten rock.

Lava—molten rock that issues from a volcano.

Magma—molten rock within the earth.

Metamorphic rocks—those changed by temperature and pressure.

Meteorite—a solid body from outer space that reaches the earth without vaporizing.

Permeability—capacity of a rock, sediment, or soil for transmitting fluids.

Porosity—the ratio of the total volume of small spaces or pores in a rock or soil to its total volume.

Salt bed—a deposit formed by the evaporation of sea water.

Sedimentary rocks—deposited in layers near the surface by water, wind, and ice.

Tectonic plate—geological concept of the movement of large segments of the earth's crust.

Volcano—a vent in the crust of the earth from which lava, gases, and ash erupt.

For nuclear terms in general, consult *Glossary of Terms in Nuclear Science and Technology,* American Nuclear Society, La Grange Park, Illinois, 1986.

For energy terms in general, consult *Energy Dictionary* by V. Daniel Hunt, Van Nostrand Reinhold, New York, 1979.

APPENDIX A
Scientific American Articles

One of the most accessible and readable sources of information on nuclear energy, energy in general, and the environment is the magazine *Scientific American*. Listed below in reverse chronological order are selected articles over recent years.

Frank von Hippel, Marvin Miller, Harold Feiveson, Anatoli Diakov, and Frans Berkhout, "Eliminating Nuclear Warheads," August 1993, p. 44.

M. Granger Morgan, "Risk Analysis and Management," July 1993, p.32.

Robert W. Conn, Valery A. Chuyanov, Nobuyuki Inoue, and Donald R. Sweetman, "The International Thermonuclear Experimental Reactor," April 1992, p. 102.

Steven Aftergood, David W. Hafemeister, Oleg F. Prilutsky, Joel R. Primack, and Stanislaus N. Rodionov, "Nuclear Power in Space," June 1991, p. 42.

Rolf F. Barth, Albert H. Solway, and Ralph G. Fairchild, "Boron Neutron Therapy for Cancer," October 1990, p. 100.

Walter Alvarez and Frank Asaro, "What Caused the Mass Extinction? An Extraterrestrial Impact," October 1990, p. 78.

John P. Holdren, "Energy in Transition," September 1990, p. 156.

Wolf Hafele, "Energy from Nuclear Power," September 1990, p. 136.

Ged R. Davis, "Energy for Planet Earth," September 1990, p. 54.

Michael W. Golay and Neil E. Todreas, "Advanced Light Water Reactors," April 1990, p. 82.

Walter Greiner and Aurel Sandulescu, "New Radioactivities," March 1990, p. 58.

Lawrence Badash, "The Age-of-the-Earth Debate," August 1989, p. 90.

Richard A. Houghton and George M. Woodwell, "Global Climatic Change," April 1989, p. 36.

Anthony R. Nero, Jr., "Controlling Indoor Air Pollution," May 1988, p. 42.

Johann Rafelski and Steven E. Jones, "Cold Nuclear Fusion," July 1987, p. 84.

William D. Phillips and Harold J. Metcalf, "Cooling and Trapping Atoms," March 1987, p. 50.

Richard P. Laeser, William I. McLaughlin, and Donna M. Wolff, "Engineering Voyager 2's Encounter with Uranus," November 1986, p. 36.

Louis Girifalco, "Materials for Energy Utilization," October 1986, p. 102.

R. Stephen Croxton, Robert L. McCrory, and John M. Soures, "Progress in Laser Fusion," August 1986, p. 68.

J. H. Hamilton and J. A. Maruhn, "Exotic Atomic Nuclei," July 1986, p. 80.

J. David Jackson, Maury Tigner, and Stanley Wojcicki, "The Superconducting Supercollider," March 1986, p. 66.

Richard K. Lester, "Rethinking Nuclear Power," March 1986, p. 31.

James A. Van Allen, "Space Science, Space Technology, and the Space Station," January 1986, p. 32.

Frank von Hippel, David H. Albright, and Barbara G. Levi, "Stopping the Production of Fissile Material for Weapons," September 1985, p. 40.

William Epstein, "A Critical Time for Nuclear Nonproliferation," August 1985, p. 33.

Leon M. Lederman, "The Value of Fundamental Science," November 1984, p. 40.

John R. Clark, "Thermal Pollution and Aquatic Life," October 1984, p.18.

Robert W. Conn, "The Engineering of Magnetic Fusion Reactors," October 1983, p. 60.

George F. Bertsch, "Vibrations of the Atomic Nucleus," May 1983, p. 62.

Edward R. Landa, "The First Nuclear Industry," November 1982, p. 180.

Arthur C. Upton, "The Biological Effects of Low-Level Ionizing Radiation," February 1982, p. 41.

Harold M. Agnew, "Gas-Cooled Nuclear Power Reactors," June 1981, p. 55.

Steven A. Fetter and Kosta Tsipis, "Catastrophic Releases of Radioactivity," April 1981, p. 41.

Wolfgang Sassin, "Energy," September 1980, p. 118.

Harold W. Lewis, "The Safety of Fission Reactors," March 1980, p. 53.

Kenneth S. Deffeyes and Ian D. MacGregor, "World Uranium Resources," January 1980, p. 66.

Robert R. Wilson, "The Next Generation of Particle Accelerators," January 1980, p. 42.

Harold P. Furth, "Progress Toward a Tokamak Fusion Reactor," August 1979, p. 50.

Gerold Yonas, "Fusion Power with Particle Beams," November 1978, p. 50.

Donald R. Olander, "The Gas Centrifuge," August 1978, p. 37.

David J. Rose and Richard K. Lester, "Nuclear Power, Nuclear Weapons, and International Stability," April 1978, p. 45.

Bernard L. Cohen, "The Disposal of Radioactive Wastes from Fission Reactors," June 1977, p. 21.

Georges Vendryes, "Superphenix: A Full-Scale Breeder Reactor," March 1977, p. 26.

Richard N. Zare, "Laser Separation of Isotopes," February 1977, p. 86.

J. D. Macdougall, "Fission-Track Dating," December 1976, p. 114.

William Bebbington, "The Reprocessing of Nuclear Fuels," December 1976, p. 30.

George A. Cowan, "A Natural Fission Reactor," July 1976, p. 36.

H. A. Bethe, "The Necessity of Fission Power," January 1976, p. 21.

Hugh C. McIntyre, "Natural-Uranium Heavy-Water Reactors," October 1975, p. 17.

Willard Bascom, "The Disposal of Waste in the Ocean," August 1974, p. 16.

John L. Emmett, John Nuckolls, and Lowell Wood, "Fusion Power by Laser Implosion," June 1974, p. 24.

R. R. Wilson, "The Batavia Accelerator," February 1974, p. 72.

David N. Schramm, "The Age of Elements," January 1974, p. 69.

M. King Hubbert, "The Energy Resources of the Earth," September 1971, p. 60.

Chauncey Starr, "Energy and Power," September 1971, p. 18.

APPENDIX B
Reference Material on Wastes

Several categories of reference material are available: (a) general reading, (b) survey papers, (c) books and reports on conferences, (d) technical articles in magazines, (e) technical reports from industry and government, and (f) bibliographies.

There is an enormous body of literature on radioactive wastes. A bibliography of the Department of Energy DOE/TIC-3311 has 13 supplements up to 1984 and contains many thousands of abstracts. A monthly bibliography from DOE is *Radioactive Waste Management.* Its categories are: High-Level Radioactive Wastes, Low-Level Radioactive Wastes, TRU Wastes, Uranium Mill Tailings, Decommissioning, Remedial Action, Waste Transport and Spent Fuel Storage, and General. Abstracts of reports, articles, books, and patents are provided.

Up-to-date information comes from conference proceedings. Some examples are listed below.

Transactions of the American Nuclear Society. This contains summary papers on research and development in all aspects of nuclear energy. It is published twice a year.

WM'94: HLW, LLW, Mixed Wastes and Environmental Restoration—Towards a Cleaner Environment, Proceedings of the Symposium on Waste Management, Tucson, Arizona, February 27-March 3, 1994. Edited by Roy G. Post, this two-volume annual collection of papers covers all aspects of radioactive waste.

International High-Level Radioactive Waste Management Conference, Las Vegas, Nevada, April 25-29, 1993. An annual event with a two-volume collection of all papers given by people from throughout the world. Sponsored

by the American Society of Civil Engineers and the American Nuclear Society, in cooperation with over 40 other organizations.

Scientific Basis for Nuclear Waste Management XVIII, recent proceedings in book form of an annual symposium of the Materials Research Society, published by that group.

We now list some of the books, reports, and articles on nuclear energy and radioactive waste management that were consulted in preparing this book or that are useful reading. Brief comments on many of the references are included.

General

Academic American Encyclopedia, Grolier, Inc., Danbury, Connecticut, 1993. Includes articles on nuclear energy and fusion energy.

Isaac Asimov, *Asimov's Biographical Encyclopedia of Science and Technology,* 2nd revised edition, Doubleday & Co., Garden City, New York, 1982. Subtitle: The Lives and Achievements of 1510 Great Scientists from Ancient Times to the Present Chronologically Arranged.

Raymond L. Murray, *Nuclear Energy,* 4th edition, Pergamon Press, Oxford and New York, 1993. A description of basic nuclear phenomena, devices, and processes, followed by a discussion of problems and opportunities. Includes applications of isotopes and radiation, nuclear reactor operation, and waste disposal. Designed for reading by college students or technically-oriented laymen.

Glossary of Terms in Nuclear Science and Technology, American Nuclear Society, La Grange Park, Illinois, 1986.

Ronald Allen Knief, *Nuclear Engineering: Theory and Technology of Commercial Nuclear Power*, 2nd Ed., Hemisphere Publishing Co., 1992. Thorough coverage at an advanced level of nuclear reactor systems, reactor safety, the nuclear fuel cycle, and nuclear fusion. An appendix is entitled, "The Impending Energy Crisis: A Perspective on the Need for Nuclear Power."

Robert E. Berlin and Catherine C. Stanton, *Radioactive Waste Management*, John Wiley & Sons, New York, 1989. Describes radioactive waste forms, regulation, processing of low-level wastes, and disposal of all kinds of waste. The last chapter deals with remediation of contaminated sites.

Y. S. Tang and James H. Saling, *Radioactive Waste Management*, Hemisphere, New York, 1990. Treats each waste type separately—spent fuel, HLW, TRU, LLW, and mill tailings. Also covers transportation and decontamination/decommissioning. Contains many tables, data, diagrams, and references.

Foo-Sun Lau, *Radioactivity and Nuclear Waste Disposal*, Research Studies Press, Letchworth, Hertfordshire, England, 1987. Section 1 contains fundamentals of radioactivity, radiation, biological effects, and radiological protection; Section 2 covers waste management, including programs around the world. Extensive tables, e.g., half-lives and radiation energies of radionuclides.

Radioactive Waste Management: An IAEA Source Book, International Atomic Energy Agency, Vienna, 1992. Designed for reading by the public, this book covers technical and institutional aspects, sociopolitical and ethical concerns, public understanding and acceptance, and organizations. It contains excellent lists of documents of the International Atomic Energy Agency and the Nuclear Energy Agency of the European coalition OECD (Organisation for Economic Co-operation and Development, based in Paris). An impressive list of contributors is included.

Fred C. Shapiro, *Radwaste, A Reporter's Investigation of a Growing Nuclear Menace*, Random House, New York, 1981. Many anecdotes about problems in waste management over the years, with reliance on quotations from opponents and officials. Generally negative conclusions.

Chapter 1—Questions and Concerns About Wastes

Charles H. Heimler and Jack Prince, *Focus on Physical Science*, Merrill Publishing Co., 1989. One of several secondary school texts that contain interesting background information. This book emphasizes investigations by the student.

Peter Alexander, et al., *Physical Science*, Silver, Burdette, and Ginn, Morristown, New Jersey, 1990. Another basic text with attractive photographs and diagrams, simple mathematics.

Robert M. Besancon, Ed., *The Encyclopedia of Physics*, 3rd edition, Van Nostrand Reinhold, New York, 1985.

David R. Lide, Ed., *CRC Handbook of Chemistry and Physics*, 74th edition, 1993-94, CRC Press, Boca Raton, Florida, 1993. Standard reference for physical data and mathematical information.

Henry N. Wagner, Jr., and Linda E. Ketchum, *Living with Radiation: The Risk, the Promise*, Johns Hopkins Press, Baltimore, 1989. Thoughtful and informative essays on radioactivity, radiation, weapons, nuclear medicine, and nuclear reactor safety.

Chapter 2—Atoms and Chemistry

Frederick E. Trinklein, *Modern Physics*, Holt, Rinehart, and Winston, Austin, Texas, 1990. A high school text in basic physics with good chapters on atomic structure, nuclear reactors, and high energy physics.

Paul Allen Tipler, *Elementary Modern Physics*, Worth Publishers, 1992. A college text in atomic and nuclear physics for students with background in basic physics.

John R. Taylor and Chris D. Zafiratos, *Modern Physics for Scientists and Engineers*, Prentice-Hall, Englewood Cliffs, New Jersey, 1991. Another excellent text in college atomic and nuclear physics.

Samuel Glasstone and Alexander Sesonske, *Nuclear Reactor Engineering*, 3rd edition, Van Nostrand Reinhold, New York, 1981.

Chapter 3—Radioactivity

W. B. Mann, R. L. Ayres, and S. B. Garfinkel, *Radioactivity and Its Measurement*, 2nd edition, Pergamon Press, Oxford, 1980.

Handbook of Radioactivity Measurement Procedures, NCRP Report No. 58, 2nd edition, National Council of Radiation Protection and Measurements, Bethesda, Maryland, 1985.

Egardo Browne and Richard Firestone; Virginia Shirley, Ed., *Table of Radioisotopes*, John Wiley & Sons, New York, 1986.

Norman Holden, "Table of the Isotopes," in *CRC Handbook of Chemistry and Physics* (See Chapter 1 references). For each isotope, gives natural abundance, atomic mass or weight, half-life, energies of radiation, and thermal neutron cross sections. Updated periodically.

Environmental Radioactivity, Proceedings of the Nineteenth Annual Meeting of the National Committee on Radiation Protection and Measurements, April 6-7, 1983, NCRP, Bethesda, Maryland, 1983. Papers are grouped: scientific, regulatory, and NCRP activities. The Taylor lecture is by Merril Eisenbud.

Robley D. Evans, *The Atomic Nucleus*, Krieger Publishers, Melbourne, Florida, 1982. A reprint of a classic 1955 advanced textbook.

Chapter 4—Kinds of Radiation

P. N. Cooper, *Introduction to Nuclear Radiation Detectors*, Cambridge University Press, 1986. A small book, easily read. No extensive knowledge of nuclear physics is assumed.

Glenn F. Knoll, *Radiation Detection and Measurement*, 2nd edition, John Wiley & Sons, New York, 1989. A comprehensive and readable college textbook.

Merril Eisenbud, *Environmental Radioactivity*, 3rd edition, Academic Press, New York, 1987. Subtitle: From Natural, Industrial, and Military Sources.

Kenneth R. Kase, Bengt E. Björngard, and Frank H. Attix, *The Dosimetry of Ionizing Radiation*, Vols. 1 (1985), 2 (1987), and 3 (1990), Academic Press, Orlando. Describes detectors and techniques for application to medical radiation physics and to accelerators.

Chapter 5—Biological Effects of Radiation

Jacob Shapiro, *Radiation Protection: A Guide for Scientists and Physicians*, 3rd edition, Harvard University Press, Cambridge, 1990. Technical but not highly mathematical; comprehensive yet practical.

Daniel S. Grosch and Larry E. Hopgood, *Biological Effects of Radiation*, 2nd edition, Academic Press, New York, 1980.

Eric J. Hall, *Radiation and Life*, 2nd edition, Pergamon Press, New York, 1984. Sources of radiation and biological effects.

Edward Pochin, *Nuclear Radiation: Risks and Benefits*, Clarendon Press, Oxford, 1983. Discusses cancer and gene damage.

Herman Cember, *Introduction to Health Physics*, Pergamon Press, New York, 1983.

James E. Turner, *Atoms, Radiation, and Radiation Protection*, Pergamon Press, New York, 1986. Text provides necessary scientific background as well as health physics information. (2nd edition, John Wiley & Sons, New York, in preparation.)

Chapter 6—Radiation Standards and Protection

Bernard Schleien and Michael S. Terpilak, *The Health Physics and Radiological Health Handbook*, Nucleon Lectern Associates, 1984. A collection of data on radiation protection, including the topics of doses, regulations, shielding, biology, and medicine. Useful information for health physicists.

Reports by the Committee on the Biological Effects of Ionizing Radiation (BEIR) of the National Research Council published by National Academy Press, Washington, D.C.:

The Effects on Populations of Exposure to Low Levels of Ionizing Radiation, 1980 (BEIR III).
Health Risks of Radon and Other Internally Deposited Alpha-Emitters—BEIR IV, 1988.
Health Effects of Exposure to Low Levels of Ionizing Radiation—BEIR V, 1990.

Chapter 7—Fission and Fission Products

Donald C. Stewart, Data for Radioactive Waste Management and Nuclear Applications, John Wiley & Sons, New York, 1985. Major sections are Physical Data (fission product properties and chains, radioisotopes, and the ORIGEN computer program); Chemical Data; Radioactive Wastes; Data for Operations.
A. Michaudon, Ed., Nuclear Fission and Neutron-Induced Fission Cross Sections, Pergamon Press, New York, 1981. An advanced book on the fission process and methods of measurement.

Chapter 8—The Manhattan Project

Robert Jungk, Brighter Than a Thousand Suns, Harcourt Brace & Co., New York, 1958. A readable history of the period 1918-1955, including the development of the first nuclear weapons.
Stephane Groueff, Manhattan Project, Little, Brown & Co., Boston, Massachusetts, 1967. Subtitle: The Untold Story of the Making of the Atomic Bomb. A highly-praised account of the period.
H. D. Smyth, Atomic Energy for Military Purposes, Princeton University Press, 1945; reprinted by AMS Press, New York, 1978 and by Stanford University Press, 1989.
Richard Rhodes, The Making of the Atomic Bomb, Simon & Schuster, New York, 1986. Thorough and well-written description of the World War II project.

Chapter 9—Defense and Development

Bertrand Goldschmidt, The Atomic Complex, American Nuclear Society, La Grange Park, Illinois, 1982. Technical and political history of nuclear weapons and nuclear power.
Stelio Villani, Isotope Separation, American Nuclear Society, La Grange Park, Illinois, 1976. Describes most of the techniques.
Richard G. Hewlett and Francis Duncan, Nuclear Navy 1946-1962, University of Chicago Press, Chicago, Illinois, 1974.
Frank G. Dawson, Nuclear Power: Development and Management of a Technology, University of Washington Press, Seattle, Washington, 1976. Covers the period 1945-1975.
Dietrich Schroeer, Science, Technology, and the Nuclear Arms Race, John Wiley & Sons, New York, 1984.
Annual Energy Review 1993, Energy Information Administration, Department of Energy, DOE/EIA-0384(93), June 1994. A wealth of statistics on sources and uses of energy of all types over many years.
T. A. Heppenheimer, "Heating Up the Cold War," Invention and Technology, Fall 1992, page 21. Reviews the history of nuclear weapon development and the nuclear Navy.
Controlled Nuclear Chain Reaction: The First Fifty Years, American Nuclear Society, La Grange Park, Illinois, 1992. Reviews the startup of the first reactor at Chicago and discusses the development of nuclear power, peaceful applications, and nuclear safety.

Chapter 10—Uses of Isotopes and Radiation

Radioisotopes: Today's Applications, DOE/NE-0089, Department of Energy. A small but informative booklet that discusses uses in medicine, science, industry, and agriculture.
Gopal B. Saha, Fundamentals of Nuclear Pharmacy, Springer-Verlag, New York, 1984. Textbook on radioactive substances used in nuclear medicine. Includes basics of radioactivity, production of radionuclides, detection of

radiation, labeling of compounds, and information on uses of radiopharmaceuticals.

Geoffrey G. Eichholz, Ed., *Radioisotope Engineering*, Marcel Dekker, New York, 1972.

Marvin Mann, *Peacetime Uses of Atomic Energy*, Thomas Y. Crowell, New York, 1975.

M. F. L'Annunziata, *Radionuclide Tracers: Their Detection and Measurement*, Academic Press, London, 1987. This book was written by a staff member of the International Atomic Energy Agency, Vienna. It emphasizes the detection process rather than the applications. Many references are listed.

G. H. Wang, David L. Willis, and Walter D. Loveland, *Radiotracer Methodology in the Biological, Environmental, and Physical Sciences*, Prentice-Hall, Englewood Cliffs, New Jersey, 1975.

John Harbert and Antonio F. G. da Rocha, *Textbook of Nuclear Medicine*, Vol. 1, Basic Science, Vol. 2, Clinical Applications, 2nd edition, Lea and Febiger, Philadelphia, Pennsylvania, 1984. A relatively readable medical book.

Industrial Applications of Radioisotopes and Radiation Technology, International Atomic Energy Agency, Vienna, 1982. Proceedings of an international conference containing many good articles.

Joseph A. Angelo, Jr. and David Buden, *Space Nuclear Power*, Orbit Book Co., Malabar, Florida, 1985. Includes isotopic power sources. (2nd edition, Krieger Publishers, Melbourne, Florida, in preparation.)

Chapter 11—Classification of Wastes

Isabelle P. Weber and Susan D. Wiltshire, *The Nuclear Waste Primer: A Handbook for Citizens*, The League of Woman Voters Education Fund, Nick Lyons Books, New York, 1985. An elementary and readable survey of the topic of radioactive wastes.

Edward L. Gershey, Robert C. Klein, Esmeralda Party, and Amy Wilkerson, *Low-Level Radioactive Waste: From Cradle to Grave*, Van Nostrand Reinhold, New York, 1990. Identifies all kinds of radioactive wastes and gives the history of commercial LLRW disposal. Covers processing and disposal. Reports on status of projects.

Chapter 12—Spent Fuel from Nuclear Reactors

Anthony V. Nero, Jr., *A Guidebook to Nuclear Reactors*, University of California Press, Berkeley, California, 1979. Excellent diagrams and photographs.

Frank J. Rahn, Achilles G. Adamantiades, John E. Kenton, and Chaim Braun, *A Guide to Nuclear Power Technology*, John Wiley & Sons, New York, 1984. A very large amount of information in the form of text, tables, diagrams, and photographs.

R. G. Wymer and B. L. Vondra, Jr., Eds., *Light Water Reactor Nuclear Fuel Cycle*, CRC Press, Boca Raton, Florida, 1981.

Chapter 13—Storage of Spent Fuel

A. M. Platt, J. V. Robinson, and O. F. Hill, *The Nuclear Fact Book*, 2nd edition, Harwood Academic Publishers, New York, 1985. A wealth of data on all aspects of energy, the nuclear fuel cycle, and nuclear wastes.

Chapter 14—Reprocessing, Recycling, and Resources

Manson Benedict, Thomas H. Pigford, and Hans Wolfgang Levi, *Nuclear Chemical Engineering*, 2nd edition, McGraw-Hill, New York, 1981.

A. J. Judd, *Fast Breeder Reactors: An Engineering Introduction*, Pergamon Press, Oxford, 1981.

Alan E. Waltar and Albert B. Reynolds, *Fast Breeder Reactors*, Pergamon Press, New York, 1981.

Uranium: 1991 Resources, Production and Demand, Organisation for Economic Co-operation and Development, Paris, 1992. The so-called "Red Book," containing tables and analyses based on estimates made by each of 40 countries.

Chapter 15—Uranium Mill Tailings

A Citizen's Guide to Radon (Second Edition): The Guide to Protecting Yourself And Your Family From Radon, U.S. Environmental Protection Agency, May 1992.

Jacob Shapiro, *Radiation Protection: A Guide for Scientists and Physicians* (3rd Ed.), Harvard University Press, Cambridge, 1990. Radiation dose calculations for radium and radon and its daughters appear on pages 165-172.

Anthony V. Nero, Jr., "A National Strategy for Indoor Radon," *Issues in Science and Technology*, Fall 1992. The article criticizes the EPA's actions and recommends a different strategy. Subsequent letters to the editor appear in the Winter 1992-1993 issue of the magazine.

Bernard L. Cohen, *Radon: A Homeowner's Guide to Detection and Control*, Consumers Union, Mount Vernon, New York, 1987. By the publishers of *Consumer Reports*. Extensive tables of data on radon concentrations by state and county, and recommendations for reducing radon in homes.

D. J. Crawford and R. W. Leggett, "Assessing the Risk of Exposure to Radioactivity," *American Scientist*, September-October 1980, p. 54. A discussion of the steps required to find the effects, using uranium mill tailings as the example source and transfer to humans via air, water, and food.

Chapter 16—Generation and Treatment of Low-Level Waste

Integrated Data Base for 1993: U.S. Spent Fuel and Radioactive Waste Inventories, Projections, and Characteristics, Oak Ridge National Laboratory, DOE/RW-0006, Rev. 9, February 1994. Extensive data on defense and commercial nuclear wastes.

A. Alan Moghissi, Herschel W. Godbee, and Sue A. Hobart, *Radioactive Waste Technology*, American Society of Mechanical Engineers, New York, 1986. Sponsored by ASME and the American Nuclear Society. Useful information on all aspects of wastes.

William Bebbington, "The Reprocessing of Nuclear Fuels," *Scientific American*,

December 1976, p. 30. This classic popular article is relevant because there have been few new technical developments.

Y. Bruce Katayama, Langdon K. Holton, Jr., Galen N. Buck, James F. Hutchens, and Mark S. Culverhouse, "Advanced Remote Decontamination Techniques Reduce Costs and Radiation Exposure," *Nuclear Technology*, July 1991, ANS. A clear description of an actual decontamination: advanced techniques are used and the results are compared with those from earlier methods.

Mark Fischetti, "When Reactors Reach Old Age," *IEEE Spectrum*, February 1986. (Institute of Electrical and Electronics Engineers.) Prospects and problems in decommissioning reactors.

G. John Weir, Jr., "Characteristics of Medically-Related Low-Level Radioactive Waste," American College of Nuclear Physicians, July 1986.

Melvin W. Carter, A. Alan Moghissi, and Bernd Kahn, *Management of Low-Level Radioactive Waste*, Volumes 1 and 2, Pergamon Press, New York and Oxford, 1979. Comprehensive technical information that is still useful despite the publication date.

Chapter 17—Transportation of Radioactive Materials

"Final Environmental Statement on the Transportation of Radioactive Material by Air and Other Modes," NUREG-0170, Nuclear Regulatory Commission, 1977. Radiation hazards due to transport of radioactive material are judged to be very small compared with those from normal transportation or from natural radiation background.

"Everything You Always Wanted to Know About Shipping High-Level Nuclear Wastes," Department of Energy, DOE/EV-0003, January 1978. Answers to 69 most often asked questions, for laymen.

Marvin Resnikoff, *The Next Nuclear Gamble*, Council on Economic Priorities, New York, 1983. Subtitle: Transportation and Storage of Nuclear Waste. By a critic and opponent of nuclear power.

Chapter 18—Health, Safety, and Environmental Protection

Samuel Glasstone and Walter H. Jordan, *Nuclear Power and Its Environmental Effects*, American Nuclear Society, La Grange Park, Illinois, 1980.

Geoffrey G. Eichholz, *Environmental Aspects of Nuclear Power*, Lewis Publishers, Chelsea, Michigan, 1985.

John E. Till and H. Robert Meyer, Eds., *Radiological Assessment, A Textbook on Environmental Dose Analysis*, Nuclear Regulatory Commission, NUREG/CR-3332, 1983. A comprehensive technical book describing techniques and models used to relate sources of radioactivity to effects on human health.

R. Allan Freeze and John A. Cherry, *Groundwater*, Prentice-Hall, Englewood Cliffs, New Jersey, 1979. A classic book on the science of water flow in geologic media and the transport of contamination.

H. C. Burkholder and E. L. J. Rosinger, "A Model for the Transport of Radionuclides and Their Decay Products Through Geologic Media," *Nuclear Technology*, Vol. 49, pp. 150-158, June 1980.

Radioactive Waste, Proceedings of the twenty-first annual meeting of the National Council on Radiation Protection and Measurements, April 3-4, 1985, NCRP, Bethesda, Maryland, 1986. Papers are grouped: scientific, regulatory, and NCRP activities. The Taylor lecture is by John H. Harley.

A. G. Milnes, *Geology and Radwaste*, Academic Press, London, 1985. A comprehensive examination of the use of the earth in radioactive waste disposal. Expresses concern about the environmental impact of radioactive wastes.

Disposal of Radioactive Waste: Review of Safety Assessment Methods, Nuclear Energy Agency, Paris, 1991. A brief summary report on performance assessment methods used on multibarrier systems in various geologic media. Describes scenario development and modeling for computation, especially for deep disposal.

Policy Implications of Greenhouse Warming, National Academy Press, Washington, D.C., 1991. A report by the Committee on Science, Engineering, and Public Policy. Provides information on the origin and nature of greenhouse gases and global warming. Considers adaptation and mitigation methods. Recommends continuing research.

Donald J. Wuebbles and Jae Edmonds, *Primer on Greenhouse Gases*, Lewis Publishers, 1991. Based on research at two institutions on the generation of gases, the relation to energy, and the effects on and of climate change. Extensive tables on the sources and properties of individual gases are presented.

Jesse H. Ausubel and Hedy E. Sladovich, Editors, *Technology and Environment*, National Academy Press, Washington, D.C., 1989. A National Academy of Engineering work on technological innovation and environmental quality. Attention is focused on improving fossil fuel power systems, with nuclear as a backstop.

Chapter 19—Disposal of Low-Level Wastes

The Shallow Land Burial of Low-Level Radioactively Contaminated Solid Waste, Committee on Radioactive Waste Management, National Academy of Sciences, Washington, D.C., 1976.

"Alternative Methods for Disposal of Low Level Radioactive Wastes," NUREG/CR3774, prepared for the Nuclear Regulatory Commission by the U.S. Army Engineer Waterways Experiment Station. Vol. 1, Task 1: Description of Methods and Assessment of Criteria, April 1984; Vol. 2, Task 2a: Belowground Vault Disposal; Vol. 3, Task 2b: Aboveground Vault; Vol. 4, Task 2c: Earth Mounded Concrete Bunker; Vol. 5, Task 2e, Shaft, all October 1985; Vol. 6, Task 2d, Mined Cavity, October 1986.

J. Howard Kittel, Editor, *Near-Surface Land Disposal*, Harwood, Chur, Switzerland, 1989. Technical information by specialists. Describes wastes, site selection, characterization, development, and operations. Options for greater-confinement disposal; environmental monitoring; corrective actions.

Directions in Low-Level Radioactive Waste Management: A Brief History of Commercial Low-Level Radioactive Waste Disposal, DOE/LLW-103, The National Low-Level Waste Management Program, Idaho Nuclear Engineering Laboratory, October 1990. Describes events and trends in three eras: Early Practices; Regulatory Reform; and State Involvement. Covers the three original commercial disposal facilities.

Marvin Resnikoff, *Living Without Landfills*, Radioactive Waste Campaign, New York, 1987. Recommends minimizing wastes, storing them at their source rather than burying them, and classifying wastes by half-life. Proposes phaseout of nuclear power.

Chapter 20—Disposal of Defense Wastes

Edward Teller, *Better a Shield than a Sword: Perspectives on Defense and Technology*, Free Press, London, 1987. "Radioactive Wastes at the Hanford Reservation, A Technical Review," National Academy of Sciences, Washington, D.C., 1978. A report prompted by concerns by outside observers about leaks from tanks. The Panel found no radiation hazard to the public but recommended remedial action. The report has served as a basis for present programs that process and dispose of defense wastes.

Department of Energy, Final Environmental Impact Statement: Disposal of Hanford Defense High-Level, Transuranic and Tank Wastes—General Summary, December 1987.

Werner Lutze and Rodney C. Ewing, *Radioactive Waste Forms for the Future*, North Holland, Amsterdam, 1988. Provides a great deal of technical information about the several forms for disposal of high-level wastes, including borosilicate glass, Synroc, and spent fuel itself. Physical and chemical properties, methods of production, and behaviors are described.

Three Department of Energy documents on Environmental Restoration and Waste Management:

1. (EM) Program, An Introduction, DOE-EM-0013P, December 1992
2. Fact Sheets

3. Five-Year Plan, Fiscal Years 1994-1998, Executive Summary, January 1993.

Chapter 21—Disposal of Spent Fuel

Science, Society, and America's Nuclear Waste, U.S. Department of Energy, DOE/RW-0361, 1992. Reading material and teacher's guides for secondary schools on the management of spent nuclear fuel and high-level waste. The Units are: 1. Nuclear Waste; 2. Ionizing Radiation; 3. The National Waste Policy Act; 4. The Waste Management System. For information phone 1(800)225-6972.

I.S. Roxburgh, *Geology of High-Level Nuclear Waste Disposal: An Introduction*, Chapman and Hall, London and New York, 1987. Thorough treatment of different media with many tables of data of chemical and physical properties. Includes a chapter on subseabed disposal.

Neil A. Chapman and Ian G. McKinley, *The Geological Disposal of Nuclear Waste*, John Wiley & Sons, Chichester, 1987. Principal attention is given to high-level waste and deep geological disposal. Release from the near-field and migration in the far-field are described. The role of models for safety assessment is reviewed.

Konrad B. Krauskopf, *Radioactive Waste Disposal and Geology*, Chapman and Hall, London and New York, 1988. Intended for readers with some knowledge of geology but not radioactive waste. Provides technical information and an identification of issues, such as disposal vs. storage or now vs. later.

Feasibility of Disposal of High-Level Radioactive Waste Into the Seabed, Volumes 1-8, Nuclear Energy Agency, Paris, 1988. A report on a ten-year study by an international Seabed Working Group, under the Organisation for Economic Co-operation and Development (OECD). Two principal emplacements—drilled holes and falling penetrometers—were considered. The method was deemed safe.

Douglas G. Brookins, *Geochemical Aspects of Radioactive Waste Disposal*, Springer-Verlag, New York, 1984. One of the few books that emphasize the chemical nature of wastes, by an expert on natural analogs.

"Report to the American Physical Society by the Study Group on Nuclear Fuel Cycle and Waste Management," *Reviews of Modern Physics*, Vol. 50, No. 1, 1978. A classic report giving a comprehensive technical evaluation of all aspects of the subject.

Management of Commercially Generated Radioactive Waste, Environmental Impact Statement, Department of Energy, DOE/EIS-0046F, 1980. Describes most of the alternative methods of disposal of high-level waste.

J. D. Bredehoeft, A. W. England, D. B. Stewart, N. J. Trask, and I. J. Winograd, "Geologic Disposal of High-Level Radioactive Wastes," U.S. Geological Survey Circular 779, Arlington, Virginia, 1978.

Ulf Lindblom and Paul Gnirk, *Nuclear Waste Disposal: Can We Rely on Bedrock?* Pergamon Press, Oxford, 1982.

Waste Isolation Systems Panel (Thomas S. Pigford et al.), *A Study of the Isolation System for Geological Disposal of Radioactive Wastes*, National Academy Press, Washington, D.C., 1984.

Luther Carter, *Nuclear Imperatives and Public Trust: Dealing With Radioactive Waste*, Resources for the Future, 1987. A thorough and interesting historical account of high-level waste and spent fuel disposal. The book is quite critical of the national programs.

"OCRWM Publications Catalog on High-Level Radioactive Waste Management," U.S. Department of Energy, Office of Civilian Radioactive Waste Management, Washington, D.C. (issued annually). References with brief abstracts on all aspects of HLW. Around 50 organizations are the sources. Citations are indexed by title, keywords, and corporate authors.

Chapter 22—Laws, Regulations, and Programs

Laws in the *United States Statutes at Large*, U.S. Government Printing Office, Washington, D.C.: "Atomic Energy Act of 1946," Public Law 585, Chapter 524, Vol. 60, Part 1, 79th Congress, 2nd Session, pp. 755-775, August 1, 1946.

"Atomic Energy Act of 1954," Public Law 703, Chapter 1073, Vol. 68, Part 1, 83rd Congress, 2nd Session, pp. 919-961, August 20, 1954.

"Low-Level Waste Policy Act" (of 1980), Public Law 99-240, Vol. 94, 96th Congress, 2nd Session, pp. 3347-3349, December 22, 1980.

"Nuclear Waste Policy Act of 1982," Public Law 97-425, 97th Congress, 2nd Session, pp . 2201-2263, January 7, 1983.

"Low-Level Radioactive Waste Amendments Act of 1985," Public Law 99-240, Vol. 99, 99th Congress, 1st Session, pp. 1842-1924, January 15, 1986.

George T. Mazuzan and J. Samuel Walker, *Controlling the Atom*, California University Press, Berkeley and Los Angeles, 1984. Subtitle: The Beginnings of Nuclear Regulation 1946-1962. A history of the regulatory role of the Atomic Energy Commission.

Code of Federal Regulations, Energy, Title 10, Parts 0-199, U.S. Government Printing Office, Washington, D.C. (annual revision). All of the NRC rules. Titles of some of the relevant sections are as follows: Part 20, Standards for Protection Against Radiation; Part 60, Disposal of High-Level Radioactive Wastes in Geologic Repositories; Part 61, Licensing Requirements for Land Disposal of Radioactive Waste; Part 71, Packaging and Transportation of Radioactive Material.

Mason Willrich and Richard K. Lester, *Radioactive Waste: Management and Regulation*, The Free Press, New York, 1977. An MIT report that emphasizes HLW and TRU in relation to nuclear power's future. Mainly of historical interest in view of changing conditions over the years.

1993 Catalog of American National Standards, American National Standards Institute, New York.

Michael E. Burns, *Low-Level Radioactive Waste Regulation: Science, Politics, and Fear*, Lewis Publishers, Chelsea, Michigan, 1988. Intended for both scientists and nonscientists. Historical background, sources of radioactive wastes, protection, siting, and political aspects.

Chapter 23—Societal Aspects of Radioactive Wastes

Bernard L. Cohen, *The Nuclear Energy Option: An Alternative for the 90s*, Plenum Press, 1990.

Mary R. English, *Siting Low-Level Radioactive Waste Disposal Facilities: The Public Policy Dilemma*, Quorum Books, New York, 1992. A description of the processes in several states, with discussion of philosophic and practical issues. Key concepts are Authority, Trust, Risk, Justice, and Legitimacy. The author does not claim to have solutions, but clearly identifies the dimensions of the low-level waste problem.

The League of Voters Education Fund, *The Nuclear Waste Primer: A Handbook for Citizens*, Lyons & Burford, Publishers, New York, 1993. Brief and readable history of waste management, its politics, the search for solutions, and the role of citizens.

E. William Colglazier, Jr., Editor, *The Politics of Nuclear Waste*, Pergamon Press, New York and Oxford, 1982. A recurring theme is that institutional issues are as important as technical issues. The dimensions of the complex interrelationships are revealed.

Nicholas Lenssen, *Nuclear Waste: The Problem That Won't Go Away*, Worldwatch Paper 106, Worldwatch Institute, Washington, D.C., December 1991. An extended editorial, critical of nuclear power and radioactive waste management.

Andrew Blowers, David Lowry, and Barry D. Solomon, *The International Politics of Nuclear Waste*, St. Martin's Press, New York, 1991. An account of waste management around the world, with emphasis on public opposition. The authors conclude that the problem is insoluble.

Ray Kemp, *The Politics of Radioactive Waste Disposal*, Manchester (UK) University Press, 1992. A thoughtful and penetrating analysis of public reactions to waste disposal facility siting.

LLW programs in the UK and Europe are emphasized.

Stanley Nealey, Barbara D. Melber, William L. Rankin, et al., *Public Opinion and Nuclear Energy*, Lexington Books, Lexington, Massachusetts, 1983. Studies by the Battelle Human Affairs Research program.

Social and Economics Aspects of Radioactive Waste Disposal, Considerations for Institutional Management, National Academy Press, Washington, D.C., 1984. A panel of the National Research Council addresses nontechnical criteria for siting geologic repositories for HLW. Transportation issues are highlighted.

Nuclear Phobia—Phobic Thinking about Nuclear Power, The Media Institute, Washington, D.C., 1980. A monograph based on discussion with Robert L. DuPont, M.D., a psychiatrist, who believes that the media accentuate people's fears.

Richard Wilson, "Analyzing the Daily Risks of Life," *Technology Review* (Massachusetts Institute of Technology), February 1979, p. 41. Semipopular comparison of risks due to air pollution, smoking tobacco, drinking alcohol, driving an auto, fire, accidents, and radiation. Shows how perceptions and reality often differ greatly.

Chapter 24—Perspectives— A Summing Up

Raymond L. Murray, "Radioactive Waste Storage and Disposal," *Proceedings of the IEEE*, Vol. 74, No. 4, April 1986, p. 552. A survey article on all types of wastes.

Nuclear Power: Technical and Institutional Options for the Future, National Academy Press, Washington, D.C., 1992. A report by the Committee on Future Nuclear Power Development of the National Research Council. Discusses nuclear utilities, advanced nuclear reactors, and R&D choices.

Index